Electromagnetic Fluctuations in Plasma

Electromagnetic Fluctuations in Plasma

Aleksei Grigor'evich

A. G. Sitenko

INSTITUT FIZIKI AKAD. NAUK, U.S.S.R.
KIEV, U.S.S.R.

Translated by Morris D. Friedman

1967

ACADEMIC PRESS   New York · London

First published in the Russian language under the title

ELEKTROMAGNITNYE FLIUKTUATSII V PLAZME

Khar'kov University Press, Khar'kov, 1965

ACADEMIC PRESS INC.
111 Fifth Avenue, New York, New York 10003

United Kingdom Edition published by
ACADEMIC PRESS INC. (LONDON) LTD.
Berkeley Square House, London W.1

LIBRARY OF CONGRESS CATALOG CARD NUMBER: 66-30103

PRINTED IN THE UNITED STATES OF AMERICA

To Lena, dear wife and comrade

Preface

This book contains a theoretical investigation of the electromagnetic properties of a plasma (an ionized gas). On the basis of the inversion of the fluctuation-dissipation theorem, a microscopic analysis of fluctuations in a plasma permits the investigation of high-frequency properties of the plasma without utilization of the kinetic equation. The high-frequency properties of both an isotropic and a magnetoactive plasma are examined in detail. Also investigated is the propagation of various kinds of waves in the plasma: electromagnetic, plasma, and hydromagnetic waves. The spectral distributions of the fluctuations of the various quantities in the plasma are determined, in particular, the case of a nonisothermal plasma is considered. The properties of a quantum and superconducting plasma are examined. Special attention is given to the electromagnetic wave scattering by fluctuations in the plasma, and also to the interaction of charged particles with the plasma. Both equilibrium and nonequilibrium plasmas are investigated, and questions of stability and instability are touched upon.

Contents

CHAPTER 1 / **General Theory of Fluctuations**

CHAPTER 2 / **Electromagnetic Fluctuations in Media
with Space-Time Dispersion**

CHAPTER 3 / **Electrodynamic Properties of an Electron
Plasma**

Electromagnetic Fluctuations
in Plasma

Introduction

At present plasma physics is one of the most vital divisions of modern physics which is experiencing a period of intensive development. The interest in plasma physics arises not only from the exceptional prevalence and variety of the phenomena in which plasma properties are manifest, but also from prospects of extensive practical utilization of plasmas in various branches of science and engineering. In the recent past, in particular, the interest in plasma physics was regenerated in connection with the unremitting search for possibilities of realizing controlled thermonuclear reactions, as well as in connection with other attempts at practical application of plasmas (development of new principles of charged-particle acceleration, creation of plasma engines, development of efficient methods of generating and amplifying microwaves, etc.).

A plasma is a fully or partially ionized gas in which electromagnetic interaction between the charged particles plays an essential role. Conventional kinetic theory, which only takes account of binary collisions between particles, is inadequate for a description of a plasma since such a theory does not describe collective effects due to long-range electromagnetic interaction between the charged particles. Plasma behavior is determined primarily by these collective effects for a sufficiently rapid change in the external fields, hence, binary collisions between particles may generally be neglected in the consideration of high-frequency properties of plasmas.

A characteristic peculiarity distinguishing a plasma from other media is the possibility of the existence therein, of weakly-damped longitudinal electrical oscillations (plasma waves), first detected by Langmuir. Later, Vlasov investigated the oscillatory properties of plasma on the basis of the kinetic equation with a self-consistent field. Landau subsequently developed a theory of plasma oscillation. In particular, Landau showed that plasma oscillations are damped even in the absence of collisions.

Another distinguishing peculiarity of plasmas is the radical change in its properties under the effect of external electric and magnetic fields. Thus, an external magnetic field leads to the origination of anisotropy in the plasma properties, and causes a number of characteristic resonant effects in the plasma. In the presence of a permanent magnetic field, the electromagnetic waves in a plasma are characterized by a number of peculiarities manifested both in the dispersion properties and in the nature of wave polarization. As has been shown by Ginzburg, these waves are of magnetohydrodynamic nature in the low-frequency ranges.

The marked peculiarities affect not only the properties of the waves being propagated in the plasma, but the character of the interaction between external charged particles and the plasma, as well. The study of the interaction between charged particles and a plasma is important for a clarification of the mechanism of exciting oscillations in plasma. The motion of charged particles at a velocity exceeding the mean thermal electron velocity in a plasma is accompanied by the excitation of oscillations in the electron density. The efficiency of the interaction increases radically when a beam of charged particles passes through the plasma. In this case the density oscillations in the system may increase exponentially with time, and the state of the system will turn out to be unstable (Akhiezer and Fainberg).

In the absence of an external magnetic field the dielectric permittivity of an unbounded plasma is less than unity, hence, Cerenkov radiation in the plasma is not possible. In the presence of an external magnetic field, the plasma refractive index is greater than one in definite frequency ranges, hence, radiation of electromagnetic waves (Cerenkov radiation) is possible during the motion of charged particles.

An investigation of the electromagnetic fluctuations in plasma yields important information on the plasma properties. According to general fluctuation theory, developed by Callen and Welton, Leontovich and Rytov, Landau and Lifshitz, electromagnetic fluctuations in a medium are determined completely by the macroscopic parameters (dielectric permittivities) characterizing the properties of the medium. Hence, if the fluctuations in the charge and current densities are determined with the aid of microscopic theory, the fluctuation–dissipation theorem may then be utilized to find the dielectric permittivities of the material. Such an approach permits taking account of thermal effects in the plasma without utilization of kinetic theory.

A nonisothermal plasma, in which the electrons and ions are characterized by a Maxwell velocity distribution but with different temperatures, possesses specific properties. The investigation of electromagnetic fluctuations in such a plasma is essential to the comprehension of the relaxation processes leading to total equilibrium.

Various nonlinear processes are possible because of the thermal fluctuations in a plasma: scattering and transformation of waves being propagated in the plasma, scattering of charged particles passing through the plasma, etc. A study of such processes may be one of the methods of determining the parameters characterizing the state of the plasma (for example, the temperature, equilibrium density, degree of nonisothermy of the plasma, etc.).

The study of electromagnetic fluctuations in a plasma also permits the direct investigation of transfer processes. The collision integral may be expressed in terms of the parameters of the electromagnetic field fluctuations because of the long-range nature of the Coulomb interaction between charged particles in plasma. In particular, knowledge of the spectral distribution of the electromagnetic field fluctuations permits the calculation of the kinetic coefficients (the coefficient of dynamic friction and the diffusion coefficients) characterizing the transfer process.

An enormous quantity of works published in various scientific journals has been devoted to the theoretical study of plasma properties. Some of them (mainly those in which the author took part) are included herein, where the electrodynamic properties of a plasma and the various processes occurring therein, particularly the electromagnetic fluctuations in a plasma, are considered. In addition to the journal papers, there are now several monographs (1–4) devoted to various questions of plasma physics, but the questions considered herein are either not treated sufficiently completely or are considered from a somewhat different viewpoint.

A list of works used in the chapter is presented at the end of each chapter. A general bibliography on the theory of fluctuations and plasma physics is presented at the end of the book.

ACKNOWLEDGMENT

I am deeply grateful to A. I. Akhiezer, with whom I jointly performed a majority of the research included in this book. I am also grateful to I. A. Akhiezer, Iu. A. Kirochkin, and V. K. Tartakovskii for assistance.

LIST OF NOTATIONS

$$\eta^2 = \frac{k^2 c^2}{\omega^2}$$

$$a^2 = \frac{T}{4\pi e^2 n_0}$$

$$\Omega^2 = \frac{4\pi e^2 n_0}{m}$$

$$s^2 = \frac{3T}{m}$$

$$t = \frac{T_e}{T_i}$$

$$\mu^2 = t\frac{M}{m}$$

$$v_s{}^2 = \frac{T_e}{M}$$

$$z = \sqrt{\frac{3}{2}}\frac{\omega}{ks}$$

$$\varphi(z) = 2ze^{-z^2}\int_0^z e^{x^2}\,dx$$

$$\omega_B = \frac{eB_0}{mc}$$

$$z_n = \sqrt{\frac{3}{2}}\frac{\omega - n\omega_B}{|k_z|s}$$

$$\beta = \frac{k^2 s^2}{3\omega_B^2}\sin^2\vartheta$$

$$\omega_\pm{}^2 = \tfrac{1}{2}(\Omega^2 + \omega_B^2)$$

$$\pm \tfrac{1}{2}\sqrt{(\Omega^2 + \omega_B^2)^2 - 4\Omega^2\omega_B^2\cos^2\vartheta}$$

$$v = \frac{\Omega^2}{\omega^2}, \qquad u = \frac{\omega_B^2}{\omega^2}$$

$$v_A = \frac{B_0}{(4\pi M n_0)^{1/2}}$$

$$\zeta = \sqrt{\frac{3}{2}\frac{v}{s}}$$

$$\Phi(\zeta) = \frac{2}{\sqrt{\pi}}\int_0^\zeta e^{-z^2}\,dz$$

$$G(\zeta) = \frac{\Phi(\zeta) - \zeta\Phi'(\zeta)}{2\zeta^2}$$

$$\mathbf{j}(\mathbf{r}, t) = \frac{1}{(2\pi)^4}\int d\mathbf{k}\,d\omega\,e^{ikr - i\omega t}\mathbf{j}(\mathbf{k}, \omega)$$

$$\mathbf{j}(\mathbf{k}, \omega) = \int d\mathbf{r}\,dt\,e^{-ikr + i\omega t}\,\mathbf{j}(\mathbf{r}, t)$$

$$\mathbf{j}_{\mathbf{k}\omega} = \mathbf{j}(\mathbf{k}, \omega)$$

$$\langle j_i j_j\rangle_{\mathbf{r}t} = \langle j_i(\mathbf{r}_1, t_1) j_j(\mathbf{r}_2, t_2)\rangle,$$
$$\mathbf{r} = \mathbf{r}_2 - \mathbf{r}_1 \quad t = t_2 - t_1$$

$$\langle j_i j_j\rangle_{\mathbf{k}\omega} = \int d\mathbf{r}\,dt\,e^{-ikr + i\omega t}\langle j_i j_j\rangle_{\mathbf{r}t}$$

$$\langle j_i j_j\rangle_{\mathbf{k}} = \frac{1}{2\pi}\int d\omega\langle j_i j_j\rangle_{\mathbf{k}\omega}$$

$$\langle j_i j_j\rangle_{\omega} = \frac{1}{(2\pi)^3}\int d\mathbf{k}\langle j_i j_j\rangle_{\mathbf{k}\omega}$$

$$\epsilon_{ij}(\omega, \mathbf{k}) = \delta_{ij} + 4\pi\chi_{ij}(\omega, \mathbf{k})$$

$$\Lambda_{ij}(\omega, \mathbf{k}) = \eta^2\left(\frac{k_i k_j}{k^2} - \delta_{ij}\right) + \epsilon_{ij}(\omega, \mathbf{k})$$

$$\Lambda(\omega, \mathbf{k}) = |\Lambda_{ij}(\omega, \mathbf{k})|$$

$$\Lambda_{ij}^0(\omega, \mathbf{k}) = \eta^2\left(\frac{k_i k_j}{k^2} - \delta_{ij}\right) + \delta_{ij}$$

$$\alpha_{ij}(\omega, \mathbf{k}) = \frac{\omega^2}{4\pi}\{\Lambda_{ij}^0 - \Lambda_{ik}^0\Lambda_{kl}^{-1}\Lambda_{lj}^0\}$$

$$\langle j_i j_j\rangle_{\mathbf{k}\omega} = \frac{\hbar}{e^{\hbar\omega/T} - 1}i\{\alpha_{ij}^*(\omega, \mathbf{k})$$
$$- \alpha_{ji}(\omega, \mathbf{k})\}.$$

IN AN ISOTROPIC MEDIUM

$$\epsilon_{ij}(\omega, \mathbf{k}) = \frac{k_i k_j}{k^2} \, \epsilon_l(\omega, \mathbf{k}) + \left(\delta_{ij} - \frac{k_i k_j}{k^2}\right) \epsilon_t(\omega, \mathbf{k}).$$

REFERENCES

1. V. L. Ginzburg, "Electromagnetic Wave Propagation in Plasmas." Fizmatgiz, Moscow, 1960 (Translations available from Gordon & Breach, New York, and Pergamon Press, New York).
2. J. G. Linhart, "Plasma Physics." North-Holland, Amsterdam, 1961.
3. L. Spitzer, "Physics of Fully Ionized Gases," 2nd ed. Wiley (Interscience), New York, 1962.
4. V. P. Silin and A. A. Rukhadze, "Electromagnetic Properties of Plasmas and Plasma-Like Media." Atomizdat, Moscow, 1961 (Gordon & Breach, New York, 1965).

I

General Theory of Fluctuations

---

1. Space-Time Correlation Functions

Let us consider the random fluctuations of some quantity distributed continuously in space. Let us denote the appropriate quantity, which we shall consider to be a vector, by $\mathbf{j}(\mathbf{r}, t)$. For definiteness, let $\mathbf{j}(\mathbf{r}, t)$ henceforth be understood to be a current in a medium. Let us assume that the selected quantity $\mathbf{j}(\mathbf{r}, t)$ is real, and that its mean value is zero in the absence of external effects. In conformity with the rules of quantum mechanics, let us form the operator $\hat{\mathbf{j}}(\mathbf{r}, t)$ of the quantity $\mathbf{j}(\mathbf{r}, t)$. Let us define the space-time Fourier components of this operator by means of the equalities

$$\mathbf{j}_{\mathbf{k}\omega} = \int d\mathbf{r} \, dt \, e^{-i\mathbf{k}r + i\omega t} \mathbf{j}(\mathbf{r}, t), \tag{1.1}$$

$$\mathbf{j}(\mathbf{r}, t) = \frac{1}{(2\pi)^4} \int d\mathbf{k} \, d\omega \, e^{i\mathbf{k}r - i\omega t} \mathbf{j}_{\mathbf{k}\omega} . \tag{1.2}$$

Correlation functions, defined as the mean values of the product of the fluctuations of some quantities at different points of space at different times, are customarily introduced for the fluctuation characteristics. The average is hence carried out on both the quantum-mechanical state of the system, and on the statistical distribution of the various quantum-mechanical states of the system.

If the medium is spatially homogeneous and stationary states of the system are considered, then the quadratic space-time correlation function will depend only on the relative distance and the absolute value of the time segment between the points at which the fluctuations are examined:

$$\langle j_i(\mathbf{r}_1, t_1) j_j(\mathbf{r}_2, t_2)\rangle = \langle j_i j_j\rangle_{\mathbf{r}t}, \tag{1.3}$$

where $\mathbf{r} = \mathbf{r}_2 - \mathbf{r}_1$ and $t = t_2 - t_1$. The brackets $\langle\cdots\rangle$ in the left-hand side of the equality denote the averaging operation. Formula (1.3) should be considered as the definition of the space-time correlation function $\langle j_i j_j\rangle_{\mathbf{r}t}$.

The spectral distribution of the space-time correlation function will be defined by using the equality

$$\langle j_i j_j\rangle_{\mathbf{k}\omega} = \int d\mathbf{r}\, dt\, e^{-i\mathbf{k}\mathbf{r}+i\omega t}\langle j_i j_j\rangle_{\mathbf{r}t}. \tag{1.4}$$

Evidently the mean value of the quadratic product of the Fourier components of the fluctuating quantities is connected to the spectral distribution of the correlation function by means of the following relationship:

$$\langle j_i{}^+(\mathbf{k}, \omega) j_j(\mathbf{k}', \omega')\rangle = (2\pi)^4\, \delta(\mathbf{k} - \mathbf{k}')\, \delta(\omega - \omega')\langle j_i j_j\rangle_{\mathbf{k}\omega}, \tag{1.5}$$

where the plus sign denotes the operation of the Hermitian conjugate.

2. Spectral Distribution of the Fluctuations and Energy Dissipation in the Medium

The spectral distribution of the correlation function for the fluctuations is determined by the dissipative properties of the medium. Let us establish a relationship between the spectral distribution of the fluctuations and the energy dissipation in the medium. We shall follow Landau and Lifshitz (1) in its exposition.

Let us first calculate the mean value of the product of $j_i{}^+(\mathbf{k}, \omega)$ and $j_j(\mathbf{k}', \omega')$. If the system is in a definite stationary state n, the quantum-mechanical mean is defined as the corresponding diagonal matrix element of the operator

$$(j_i{}^+(\mathbf{k}, \omega) j_j(\mathbf{k}', \omega'))_{nn} = \sum_m j_i{}^+(\mathbf{k}, \omega)_{nm} j_j(\mathbf{k}', \omega')_{mn}. \tag{1.6}$$

[In (1.6) the summation extends over all states of the system.] Evaluating the matrix elements of the operator $j_{k\omega}$ by using wavefunctions of the stationary states, we find

$$(\mathbf{j}_{k\omega})_{nm} = 2\pi\delta(\omega + \omega_{nm})(\mathbf{j}_k)_{nm}, \tag{1.7}$$

where $\omega_{nm} = (E_n - E_m)/\hbar$ is the frequency of the transition from the stationary state n to the stationary state m, and $(\mathbf{j}_k)_{nm}$ is the matrix element of the operator $\mathbf{j}_k$, independent of the time. Substituting (1.7), as well as an analogous expression for $(\mathbf{j}_{k\omega}^+)_{mn}$, into (1.6) and taking the statistical average, we have

$$\langle j_i^+(\mathbf{k}, \omega) j_j(\mathbf{k}', \omega')\rangle = 2\pi \, \delta(\omega - \omega')\langle j_i^+(\mathbf{k}) j_j(\mathbf{k}')\rangle_\omega, \tag{1.8}$$

$$\langle j_i^+(\mathbf{k}) j_j(\mathbf{k}')\rangle_\omega = 2\pi \sum_{m,n} f(E_n) \, j_i^+(\mathbf{k})_{nm} j_j(\mathbf{k}')_{mn} \, \delta(\omega - \omega_{nm}), \tag{1.9}$$

where $f(E_n)$ is the statistical distribution function of the various quantum-mechanical states of the system. In the case of an equilibrium distribution of the states of the system, the function $f(E_n)$ is defined by the Gibbs distribution

$$f(E_n) = \exp(F - E_n)/T, \tag{1.10}$$

where F is the free energy and T is the absolute temperature of the system.

The correlation function (1.9) may be connected directly to the energy absorbed by the system as a result of dissipation. Let us assume that a periodic perturbation with frequency ω, whose energy is proportional to the quantity $\mathbf{j}$, acts on the system. Selecting the potential $\mathbf{A}(\mathbf{r}, t)$ as the perturbation, the energy operator of the perturbation may be written as

$$V = - \int d\mathbf{r} \, \mathbf{A}(\mathbf{r}, t) \, \mathbf{j}(\mathbf{r}, t). \tag{1.11}$$

Transforming to Fourier components in (1.11) and using the complex form of writing the quantities $\mathbf{A}$ and $\mathbf{j}$, we represent the perturbation energy as

$$V = - \tfrac{1}{2} \operatorname{Re} \sum_{\mathbf{k}} \mathbf{A}_{\mathbf{k}}(t) \, j_{\mathbf{k}}^+(t), \tag{1.12}$$

where $A_k(t)$ is a harmonic function of the time

$$A_k(t) = A_{k\omega} e^{-i\omega t}.$$

Transitions between different states of the system are possible under the influence of the perturbation (1.11). Using (1.12) it is easy to evaluate the matrix element of the transition of the system $n \to m$:

$$V_{nm} = -\pi \sum_k \{A_{k\omega}(j_k^+)_{nm}\, \delta(\omega - \omega_{nm}) + A_{k\omega}^*(j_k)_{nm}\, \delta(\omega + \omega_{nm})\}. \quad (1.13)$$

By using (1.13) we obtain the following expression for the transition probability of the system per unit time:

$$w_{nm} = \frac{\pi}{2\hbar^2} \sum_{k,k'} A_i(\mathbf{k},\, \omega)\, A_j^*(\mathbf{k}',\, \omega)\{j_i^+(\mathbf{k})_{mn} j_j(\mathbf{k}')_{nm}\, \delta(\omega + \omega_{nm})$$

$$+ j_i^+(\mathbf{k})_{nm} j_j(\mathbf{k}')_{mn}\, \delta(\omega - \omega_{nm})\}. \quad (1.14)$$

With each transition $n \to m$ the system absorbs the energy quantum $\hbar\omega_{mn}$, whose source is the external perturbation. Hence, the energy absorbed per unit time equals

$$Q_n = \sum_m w_{nm}\hbar\omega_{mn}. \quad (1.15)$$

We find the mean value of the absorbed energy by averaging (1.15) over all stationary states n:

$$Q = \sum_{m,n} f(E_n)w_{nm}\hbar\omega_{mn}. \quad (1.16)$$

Substituting (1.14) into (1.16) and interchanging the summation indices n and m in one of the members, we find

$$Q = \frac{\pi\omega}{2\hbar} \sum_{k,k'} A_i(\mathbf{k},\, \omega)A_j^*(\mathbf{k}',\, \omega) \sum_{m,n} \{f(E_n - \hbar\omega) - f(E_n)\}$$

$$\times j_i^+(\mathbf{k})_{nm} j_j(\mathbf{k}')_{mn}\, \delta(\omega - \omega_{nm}). \quad (1.17)$$

Comparing this expression with (1.9) we establish the following relationship, connecting the mean energy absorbed by the system per unit time to the correlation function of the fluctuating quantities:

$$Q = \frac{\omega}{4\hbar} \sum_{\mathbf{k}, \mathbf{k}'} A_i(\mathbf{k}, \omega) A_j{}^*(\mathbf{k}', \omega)\{\langle j_i{}^+(\mathbf{k}) j_j(\mathbf{k}')\rangle_\omega^{\hbar\omega}$$

$$- \langle j_i{}^+(\mathbf{k}) j_j(\mathbf{k}')\rangle_\omega\}, \tag{1.18}$$

where $\langle j_i{}^+(\mathbf{k}) j_j(\mathbf{k}')\rangle_\omega^{\hbar\omega}$ is the correlation function in which the averaging is over the energy distribution displaced by the quantity $\hbar\omega$:

$$\langle j_i{}^+(\mathbf{k}) j_j(\mathbf{k}')\rangle_\omega^{\hbar\omega} = 2\pi \sum_{m,n} f(E_n - \hbar\omega) j_i{}^+(\mathbf{k})_{nm} j_j(\mathbf{k}')_{mn}\, \delta(\omega - \omega_{nm}). \tag{1.9'}$$

On the other hand, the absorbed energy Q may be connected with the macroscopic parameters characterizing the dissipative properties of the system.

In the absence of external perturbations, the mean value of the quantity $\mathbf{j}$ is zero. Under the effect of the external perturbation (1.11), a nonzero value of $\mathbf{j}$, proportional to the magnitude of the perturbing potential $\mathbf{A}$, occurs

$$\langle j_i \rangle = \hat{\alpha}_{ij} A_j, \tag{1.19}$$

where $\hat{\alpha}_{ij}$ is the linear space-time integral operator. Using the complex form of writing $\mathbf{j}$ and $\mathbf{A}$, this linear relation may be represented as

$$j_i(\mathbf{k}, \omega) = \alpha_{ij}(\omega, \mathbf{k}) A_j(\mathbf{k}, \omega),$$

where the $\alpha_{ij}(\omega, \mathbf{k})$ are macroscopic coefficients characterizing the dissipative properties of the system. Let us note that the relationship (1.19) has meaning only for an unbounded spatially homogeneous medium.

The processes for which (1.19) holds are customarily called linear dissipative processes. In the case of a linear dissipative process the absorbed energy Q is expressed directly in terms of the coefficients α_{ij}. Indeed, a change in the mean internal energy of the system equals the mean value of the partial derivative of the system Hamiltonian with respect to the time. Since only the perturbation V depends explicitly on the time in the Hamiltonian, we then have for the change in the internal energy of the system

$$\frac{\partial U}{\partial t} = -\int d\mathbf{r}\, \dot{\mathbf{A}}\,(\mathbf{r}, t) \langle \mathbf{j}(\mathbf{r}, t)\rangle. \tag{1.20}$$

We find the mean energy absorbed per unit time by averaging (1.20) with respect to the period of the external perturbation. Transforming to Fourier components in (1.20) and using (1.19), we obtain after taking the average over the period

$$Q = \frac{\omega}{4} i \sum_{\mathbf{k}} (\alpha_{ij}^* - \alpha_{ji}) A_i(\mathbf{k}, \omega) A_j^*(\mathbf{k}, \omega). \qquad (1.21)$$

Comparing this expression with (1.18) and taking account of (1.8) and (1.5) we find

$$\langle j_i j_j \rangle_{\mathbf{k}\omega}^{\hbar\omega} - \langle j_i j_j \rangle_{\mathbf{k}\omega} = \hbar i \{\alpha_{ij}^*(\omega, \mathbf{k}) - \alpha_{ji}(\omega, \mathbf{k})\}. \qquad (1.22)$$

This relationship establishes the general connection between the correlation function of the fluctuating quantities and the dissipative properties of the system characterized by the coefficients α_{ij}. Formula (1.22) is valid both for thermodynamically equilibrium quantities and for the general case of systems in nonequilibrium states.

In the classical case, (1.22) may be simplified by using the expansion of the distribution function $f(E_n - \hbar\omega)$ in $\langle j_i j_j \rangle_{\mathbf{k}\omega}^{\hbar\omega}$ in a power series of the transferred energy $\hbar\omega$. In the limit as $\hbar \to 0$ we have

$$\omega \frac{\partial}{\partial E} \langle j_i j_j \rangle_{\mathbf{k}\omega} = i\{\alpha_{ji}(\omega, \mathbf{k}) - \alpha_{ij}^*(\omega, \mathbf{k})\}, \qquad (1.23)$$

where

$$\frac{\partial}{\partial E} \langle j_i j_j \rangle_{\mathbf{k}\omega} \delta(\mathbf{k} - \mathbf{k}') \equiv (2\pi)^{-2} \sum_{m,n} \frac{\partial f(E_n)}{\partial E_n} j_i^+(\mathbf{k})_{nm} j_j(\mathbf{k}')_{mn} \delta(\omega - \omega_{nm}).$$

If the system under consideration is a set of independent subsystems, each of which is characterized by its energy E^α, and the transferred energy $\hbar\omega$ is distributed between the separate subsystems, then (1.23) should be modified; namely, the substitution

$$\omega \frac{\partial}{\partial E} \langle j_i j_j \rangle_{\mathbf{k}\omega} \to \sum_{\alpha} \omega_\alpha \frac{\partial}{\partial E^\alpha} \langle j_i j_j \rangle_{\mathbf{k}\omega}, \qquad (1.24)$$

should be made in the left side of equality (1.23), where $\hbar\omega_\alpha$ is the change in energy in the individual subsystem, and $\sum_\alpha \omega_\alpha = \omega$.

Limiting ourselves henceforth to the consideration of equilibrium distributions of systems, let us use the Gibbs distribution (1.10) as $f(E_n)$. In this case, taking account of the multiplicative character of the correlations functions with the energy shift

$$\langle j_i j_j \rangle_{\mathbf{k}\omega}^{\hbar\omega} = e^{\hbar\omega/T} \langle j_i j_j \rangle_{\mathbf{k}\omega},$$

we obtain the following formula from (1.22) for the spectral distribution of the correlation function of the fluctuations:

$$\langle j_i j_j \rangle_{\mathbf{k}\omega} = \frac{\hbar}{\exp(\hbar\omega/T) - 1} i(\alpha_{ij}^*(\omega, \mathbf{k}) - \alpha_{ji}(\omega, \mathbf{k})\}. \qquad (1.25)$$

This formula, which is a generalization of the Nyquist fluctuation–dissipative theorem (2), completely defines the fluctuations of the distributed quantities in an equilibrium system.*

Having selected the quantity j for a specific process, and knowing the explicit expression for the change in energy, it is not difficult to determine what quantity plays the part of the appropriate **A** by comparing the expression for the change in energy with (1.20). Then finding the coefficients α_{ij} from the equations describing the motion of the system, we determine completely the fluctuations of the quantities of interest to us according to (1.25). Formula (1.25) may be modified somewhat by considering that the fluctuations of **j** are a result of the effect of some random fictional potential **A**. Remarking that

$$A_i = \alpha_{ij}^{-1} j_j, \qquad (1.26)$$

by using (1.25) we find for the spectral distribution of the mean square of the random potential

$$\langle A_i A_j \rangle_{\mathbf{k}\omega} = \frac{\hbar}{\exp(\hbar\omega/T) - 1} i\{\alpha_{ji}^{-1}(\omega, \mathbf{k}) - \alpha_{ij}^{-1*}(\omega, \mathbf{k})\}. \qquad (1.27)$$

The use of (1.27) in place of (1.25) has definite advantages for a number of specific applications of the theory.

* The expounded theory of fluctuations has been developed by Callen and Welton (3), Leontovich and Rytov (4, 5), and Landau and Lifshitz (1). Electromagnetic fluctuations in media with spatial dispersion have been analyzed by Silin (6) [see also Silin and Rukhadze (7)].

Let us present the definition of a symmetrized space-time correlation function

$$\langle j_i j_j \rangle^s_{rt} = \tfrac{1}{2}\langle j_i(\mathbf{r}_1, t_1) j_j(\mathbf{r}_2, t_2) + j_j(\mathbf{r}_2, t_2) j_i(\mathbf{r}_1, t_1) \rangle. \qquad (1.28)$$

As is easy to verify, the spectral distribution of the symmetrized correlation function is determined by the expression

$$\langle j_i j_j \rangle^s_{\mathbf{k}\omega} = \frac{\hbar}{2} \, \text{cotanh} \, \frac{\hbar\omega}{T} \, i\{\alpha^*_{ij}(\omega, \mathbf{k}) - \alpha_{ji}(\omega, \mathbf{k})\}. \qquad (1.29)$$

For sufficiently high temperatures satisfying the condition $T \gg \hbar\omega$, both (1.25) and (1.29) take the form

$$\langle j_i j_j \rangle_{\mathbf{k}\omega} = \frac{T}{\omega} \, i\{\alpha^*_{ij}(\omega, \mathbf{k}) - \alpha_{ji}(\omega, \mathbf{k})\}, \qquad (1.30)$$

that is, the spectral distribution of the fluctuations turns out to be independent of the quantum constant $\hbar$, in conformity with the fact that the fluctuations are classical under these conditions.

3. Properties of the Tensor of the Linear Relating Coefficients α_i

The tensor of the coefficients $\alpha_{ij}(\omega, \mathbf{k})$, which establish a linear relation between the quantities $j_i(\mathbf{k}, \omega)$ and $A_j(\mathbf{k}, \omega)$, is a complex function of the real variables ω and $\mathbf{k}$ in the general case. Let α'_{ij} and α''_{ij} denote the real and imaginary parts of this tensor:

$$\alpha_{ij} = \alpha'_{ij} + i\alpha''_{ij}.$$

It is easy to establish general properties of the tensor $\alpha_{ij}(\omega, \mathbf{k})$. From the condition that the quantity $j(\mathbf{r}, t)$ $(j_{\mathbf{k}\omega} = j^+_{-\mathbf{k}-\omega})$ be real for a given real value of $\mathbf{A}(\mathbf{r}, t)$ $(\mathbf{A}_{\mathbf{k}\omega} = \mathbf{A}^+_{-\mathbf{k}-\omega})$, it follows that the coefficients $\alpha_{ij}(\omega, \mathbf{k})$ should satisfy the relationship

$$\alpha_{ij}(\omega, \mathbf{k}) = \alpha^*_{ij}(-\omega, -\mathbf{k}). \qquad (1.31)$$

Separating real and imaginary parts in (1.31), we find

$$\alpha'_{ij}(\omega, \mathbf{k}) = \alpha'_{ij}(-\omega, -\mathbf{k}), \qquad \alpha''_{ij}(\omega, \mathbf{k}) = -\alpha''_{ij}(-\omega, -\mathbf{k}). \qquad (1.32)$$

The symmetry of the tensor $\alpha_{ij}(\omega, \mathbf{k})$ relative to the subscripts i and j may be determined by taking account of the invariance of the quadratic correlation function (1.28) relative to changes in the sign of the time. Since a change in the sign of the time is equivalent to replacement of ω by $-\omega$, in the Fourier component, then

$$\langle j_i j_j \rangle^s_{\mathbf{k}\omega} = \langle j_i j_j \rangle^s_{\mathbf{k}-\omega}.$$

Using the definition of the symmetrized correlation function (1.28) it is easy to verify that

$$\langle j_i j_j \rangle^s_{\mathbf{k}-\omega} = \langle j_j j_i \rangle^s_{-\mathbf{k}\omega},$$

from which results the following symmetry property of the tensor:

$$a_{ij}(\omega, \mathbf{k}) = \alpha_{ji}(\omega, -\mathbf{k}). \tag{1.33}$$

The relationship (1.33) is modified somewhat if the system is in an external magnetic field $\mathbf{B}_0$. In this case $\mathbf{B}_0$ should simultaneously be replaced by $-\mathbf{B}_0$ when the sign of the time is changed. Hence, we obtain in place of (1.33)

$$\alpha_{ij}(\omega, \mathbf{k}, \mathbf{B}_0) = \alpha_{ji}(\omega, -\mathbf{k}, -\mathbf{B}_0). \tag{1.34}$$

The relationships (1.33) and (1.31) simplify substantially in the case of an isotropic medium in the absence of external constant fields. In fact, since there is a single vector $\mathbf{k}$ in the isotropic case, on which $\alpha_{ij}(\omega, \mathbf{k})$ depends, the tensor $\alpha_{ij}(\omega, \mathbf{k})$ may be represented as a linear combination of uniquely possible tensors of the second rank δ_{ij} and $k_i k_j$. The coefficients for these tensors may depend only on the modulus of the vector $\mathbf{k}$. It hence follows at once that in the isotropic case the tensor $\alpha_{ij}(\omega, \mathbf{k})$ is symmetric relative to the subscripts i and j

$$\alpha_{ij}(\omega, \mathbf{k}) = \alpha_{ji}(\omega, \mathbf{k}) \tag{1.35}$$

and is an even function in the vector $\mathbf{k}$

$$\alpha_{ij}(\omega, \mathbf{k}) = \alpha_{ij}(\omega, -\mathbf{k}). \tag{1.36}$$

Because of the evenness of $\alpha_{ij}(\omega, \mathbf{k})$ in $\mathbf{k}$, the relationships (1.32) will define the frequency dependence of the tensor $\alpha_{ij}(\omega, \mathbf{k})$. According to (1.32), in this case the real part of the tensor $\alpha_{ij}(\omega, \mathbf{k})$

is an even function of the frequency, and the imaginary part of $\alpha_{ij}(\omega, \mathbf{k})$ is an odd function:

$$\alpha'_{ij}(\omega, \mathbf{k}) = \alpha'_{ij}(-\omega, \mathbf{k}), \qquad \alpha''_{ij}(\omega, \mathbf{k}) = -\alpha''_{ij}(-\omega, \mathbf{k}). \qquad (1.37)$$

The tensor $\alpha_{ij}(\omega, \mathbf{k})$, considered as a function of the complex variable ω, is an analytic function in the upper ω half-plane. In fact, since the value of $\mathbf{j}(\mathbf{r}, t)$ at some time t may depend only on the external effect $\mathbf{A}(\mathbf{r}, t')$ at preceding times t', the coefficient $\alpha_{ij}(\omega, \mathbf{k})$ may not have poles in the upper complex ω half-plane. From the connection between the imaginary part of $\alpha_{ij}(\omega, \mathbf{k})$ with the energy dissipation (1.20) it also follows that $\alpha''_{ij}(\omega, \mathbf{k})$ has no zeroes for real, nonzero values of ω.

A number of other properties may be derived from these properties of the tensor $\alpha_{ij}(\omega, \mathbf{k})$ as a function of the complex frequency ω. In particular, it is possible to obtain the Kramers–Kronig relationship establishing a connection between the real and imaginary parts of the tensor $\alpha_{ij}(\omega, \mathbf{k})$:

$$\alpha'_{ij}(\omega, \mathbf{k}) - \alpha_{ij}(\infty, \mathbf{k}) = \frac{1}{\pi} \int_{-\infty}^{\infty} \frac{\alpha''_{ij}(\omega', \mathbf{k})}{\omega' - \omega} \, d\omega', \qquad (1.38)$$

$$\alpha''_{ij}(\omega, \mathbf{k}) = \frac{1}{\pi} \int_{-\infty}^{\infty} \frac{\alpha_{ij}(\infty, \mathbf{k}) - \alpha'_{ij}(\omega', \mathbf{k})}{\omega' - \omega} \, d\omega'. \qquad (1.39)$$

Here $\alpha_{ij}(\infty, \mathbf{k})$ is the value of the tensor $\alpha_{ij}(\omega, \mathbf{k})$ at $\omega = \infty$.

4. Two-Particle Green's Function

Two-particle Green's functions may be used to describe the fluctuations in place of correlation functions. The Green's functions have the advantage over correlation functions that their spectral distributions admit analytic continuation in the complex plane.

The two-particle Green's function for the current (the Green's tensor) may be defined as follows. The causal Green's function is defined as the mean value of the chronological product of current operators [Schwinger (8)]:

$$G^c_{ij}(\mathbf{r}_2 - \mathbf{r}_1, t_2 - t_1) = -i\langle T j_i(\mathbf{r}_1, t_1) j_i(\mathbf{r}_2, t_2)\rangle. \qquad (1.40)$$

Here the symbol T denotes the chronological or T-product of the current operators

$$
\begin{aligned}
Tj_j(\mathbf{r}_1, t_1) j_i(\mathbf{r}_2, t_2) &= \theta(t_1 - t_2) j_j(\mathbf{r}_1, t_1) j_i(\mathbf{r}_2, t_2) \\
&+ \theta(t_2 - t_1) j_i(\mathbf{r}_2, t_2) j_j(\mathbf{r}_1, t_1),
\end{aligned} \tag{1.41}
$$

where $\theta(t)$ is the discontinuous Heaviside function

$$
\theta(t) = \begin{matrix} 0, & t < 0, \\ 1, & t > 0. \end{matrix} \tag{1.42}
$$

However, it is more convenient to use the retarding or advancing Green's functions [Bogoliubov and Tiablikov (9)] defined by means of the equalities

$$
G_{ij}^R(\mathbf{r}_2 - \mathbf{r}_1, t_2 - t_1) = -i\theta(t_1 - t_2)\langle[j_j(\mathbf{r}_1, t_1), j_i(\mathbf{r}_2, t_2)]\rangle, \tag{1.43}
$$

$$
G_{ij}^A(\mathbf{r}_2 - \mathbf{r}_1, t_2 - t_1) = i\theta(t_2 - t_1)\langle[j_j(\mathbf{r}_1, t_1), j_i(\mathbf{r}_2, t_2)]\rangle. \tag{1.44}
$$

The square brackets in (1.43) and (1.44) denote the commutator $[a, b] = ab - ba$. Let us recall that the brackets $\langle \cdots \rangle$ in (1.40), (1.43), and (1.44) denote the statistical averaging operation. The spectral representations for the retarding and advancing Green's functions may be connected to the spectral representation of the correlation function. Taking the Fourier transformation in (1.43) and remarking that

$$
\langle j_j j_i \rangle_{-\mathbf{k}-\omega} = \exp\left(\frac{\hbar\omega}{T}\right) \langle j_i j_j \rangle_{\mathbf{k}\omega},
$$

we have

$$
G_{ij}^R(\mathbf{k}, \omega) = \frac{i}{2\pi} \int_{-\infty}^{\infty} d\omega' \left(1 - \exp\left[\frac{\hbar\omega'}{T}\right]\right) \langle j_i j_j \rangle_{\mathbf{k}\omega'} \int_{-\infty}^{\infty} dt\, e^{i(\omega-\omega')t}\theta(t). \tag{1.45}
$$

The discontinuous Heaviside factor $\theta(t)$ may be represented as the integral

$$
\theta(t) = \frac{i}{2\pi} \int_{-\infty}^{\infty} \frac{e^{-it\xi}}{\xi + io} d\xi. \tag{1.46}
$$

Considering ξ as a complex variable, the integration in (1.46) may be performed along the contour shown in Fig. 1. The integrand

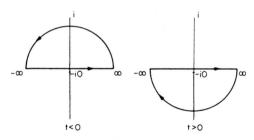

$t<0$ $t>0$

FIG. 1.

has a pole at $\xi = -io$ in the lower half-plane. If $t < 0$, the contour should be closed in the upper half-plane, and the integral equals zero. If $t > 0$, the contour should be closed in the lower half-plane, and the integral equals one. Using (1.46) we have

$$\int_{-\infty}^{\infty} dt \, e^{i(\omega - \omega')t}\theta(t) = \frac{i}{\omega - \omega' + io}. \tag{1.47}$$

Hence, we obtain the following formula for the spectral representation of the retarding Green's function:

$$G_{ij}^R(\mathbf{k}, \omega) = \frac{1}{2\pi} \int_{-\infty}^{\infty} d\omega' \left(1 - \exp\left[\frac{\hbar\omega'}{T}\right]\right) \frac{\langle j_i j_j \rangle_{\mathbf{k}\omega'}}{\omega' - \omega - io}. \tag{1.48}$$

The advancing Green's function $G_{ij}^A(\mathbf{k}, \omega)$ is defined by the same formula with $-io$ replaced by io.

The function (1.48) may be continued analytically in the complex ω plane. Indeed, for complex ω we have

$$\frac{1}{2\pi} \int_{-\infty}^{\infty} d\omega' \left(1 - \exp\left[\frac{\hbar\omega'}{T}\right]\right) \frac{\langle j_i j_j \rangle_{\mathbf{k}\omega'}}{\omega' - \omega} = \begin{cases} G_{ij}^R(\mathbf{k}, \omega), & \text{Im } \omega > 0, \\ G_{ij}^A(\mathbf{k}, \omega), & \text{Im } \omega < 0. \end{cases} \tag{1.49}$$

Hence, $G_{ij}^{R,A}(\mathbf{k}, \omega)$ may be considered as a single analytic function $G_{ij}(\mathbf{k}, \omega)$ in the complex ω plane, which has a slit on the real axis. If the Green's function $G_{ij}(\mathbf{k}, \omega)$ is known, the spectral distribution of the correlation function is determined by the relationship

$$\langle j_i j_j \rangle_{\mathbf{k}\omega} = \frac{i}{\exp[\hbar\omega/T] - 1} \{G_{ij}(\mathbf{k}, \omega + io) - G_{ij}(\mathbf{k}, \omega - io)\}. \tag{1.50}$$

Using the equation of motion for the current operator j_i, a system of linking equations may be obtained for the Green's functions. A higher order Green's function than the original one will hence enter in the right-hand side of each equation. If a finite chain can be broken off by using some kind of approximate method, i.e., if a finite system of equations can be obtained, then finding the Green's functions reduces to solving such a system. Just as the correlation functions, the two-particle Green's functions describe the properties of systems of interacting particles sufficiently completely. As we will see later, the dielectric permittivity tensor of a material is expressed directly in terms of the retarding two-particle Green's function for the currents.

REFERENCES

1. L. D. Landau and E. M. Lifshitz, "Electrodynamics of Continuous Media," Chapter 13. Gostekhizdat, Moscow, 1957 (English translation, Addison-Wesley, Reading, Massachussetts, 1958).
2. H. Nyquist, *Phys. Rev.* **32**, 110 (1928).
3. H. Callen and T. Welton, *Phys. Rev.* **83**, 34 (1951).
4. M. A. Leontovich and S. M. Rytov, *ZETF* **23**, 246 (1952).
5. S. M. Rytov, "Theory of Electrical Fluctuations and Thermal Radiation." AN USSR Press, Moscow, 1953 (English translation, AFCRL TR 59-162).
6. V. P. Silin, *Izv. VUZ, Radiofizika* **2**, 98 (1959) (English translation available from CFSTI [Center for Federal Scientific and Technical Information]).
7. V. P. Silin and A. A. Rukhadze, "Electromagnetic Properties of Plasmas and Plasma-Like Media," Sec. 9. Atomizdat, Moscow, 1961 (English translation, Gordon & Breach, New York, 1965).
8. J. Schwinger, *Proc. Natl. Acad. Sci.* **37**, 452, 455 (1951).
9. N. N. Bogoliubov and S. V. Tiablikov, *DAN SSSR* **126**, 53 (1959) (English translation, see *Sov. Phys. Dokl.*).

2

Electromagnetic Fluctuations in Media with Space-Time Dispersion

1. Dielectric Permittivity Tensor

Let us consider an unbounded spatially homogeneous medium possessing space-time dispersion in the general case. The macroscopic electromagnetic field in such a medium is defined by the system of Maxwell equations

$$\text{rot } \mathbf{E} = -\frac{1}{c}\frac{\partial \mathbf{B}}{\partial t},$$

$$\text{div } \mathbf{B} = 0,$$

$$\text{rot } \mathbf{B} = \frac{1}{c}\frac{\partial \mathbf{E}}{\partial t} + \frac{4\pi}{c}\mathbf{j}, \qquad (2.1)$$

$$\text{div } \mathbf{E} = 4\pi\rho,$$

where ρ and $\mathbf{j}$ are the densities of the induced charge and the currents which originate in the medium under the effect of the electromagnetic field. (For simplicity, we consider the field in the absence of free charges and currents.) The densities of the induced charges and currents are interrelated by means of the continuity equation

$$\frac{\partial \rho}{\partial t} + \text{div } \mathbf{j} = 0. \qquad (2.2)$$

The system of equations (2.1) should be supplemented by a relationship connecting the induced current density with the electric field intensity, which is customarily called the material equation. The presence of time dispersion in the medium denotes that the induced charges and currents depend at a given time on the values of the field in all the preceding times; the presence of spatial dispersion means, in addition, that there is a nonlocal connection between the mentioned quantities. [Gertsenshtein (1) first introduced the terminology spatial dispersion.] The field of electrical induction $\mathbf{D}$, defined by using the equality (2)

$$\frac{\partial \mathbf{D}}{\partial t} = \frac{\partial \mathbf{E}}{dt} + 4\pi \mathbf{j} \tag{2.3}$$

may be inserted into the system (2.1) in place of the induced charges and currents. By using the continuity equation (2.2), the system (2.1) may hence be rewritten as follows:

$$\operatorname{rot} \mathbf{E} = -\frac{1}{c}\frac{\partial \mathbf{B}}{\partial t}, \qquad \operatorname{div} \mathbf{B} = 0,$$

$$\operatorname{rot} \mathbf{B} = \frac{1}{c}\frac{\partial \mathbf{D}}{\partial t}, \qquad \operatorname{div} \mathbf{D} = 0. \tag{2.4}$$

The relationship (2.3) establishing a connection between the induction and the intensity of the electric field by taking account of the material equation is written formally for linear processes as

$$D_i = \hat{\epsilon}_{ij} E_j, \tag{2.5}$$

where $\hat{\epsilon}_{ij}$ is the dielectric permittivity tensor characterizing the electromagnetic properties of the medium. The tensor $\hat{\epsilon}_{ij}$ is a linear space-time integral operator. Representing the fields $\mathbf{E}$ and $\mathbf{D}$ as space-time Fourier integrals, we obtain the following relationship for the Fourier component from (2.5):

$$D_i(\mathbf{k}, \omega) = \epsilon_{ij}(\omega, \mathbf{k})\, E_j(\mathbf{k}, \omega). \tag{2.6}$$

In the general case of a medium with space-time dispersion, the components of the dielectric permittivity tensor turn out to be dependent on both the frequency ω and the wave vector $\mathbf{k}$.

Since the definition (2.6) is completely analogous to the definition of the linear relating coefficients (1.19), the properties of the dielectric permittivity tensor $\epsilon_{ij}(\omega, \mathbf{k})$ are then also completely similar to the properties of the tensor $\alpha_{ij}(\omega, \mathbf{k})$. The structure of the tensor $\epsilon_{ij}(\omega, \mathbf{k})$ is simplest in the case of an isotropic medium in the absence of constant external fields. Since the tensor ϵ_{ij} depends in this case only on the single vector $\mathbf{k}$, it may then be represented as a linear combination of two tensor quantities δ_{ij} and $k_i k_j$. It is convenient to select this combination in the form

$$\epsilon_{ij}(\omega, \mathbf{k}) = \frac{k_i k_j}{k^2} \epsilon_l(\omega, \mathbf{k}) + \left(\delta_{ij} - \frac{k_i k_j}{k^2} \right) \epsilon_t(\omega, \mathbf{k}). \tag{2.7}$$

The coefficients ϵ_l and ϵ_t may be considered here as the longitudinal and transverse components of the dielectric permittivity.

The electrical susceptibility tensor χ_{ij}

$$\epsilon_{ij}(\omega, \mathbf{k}) = \delta_{ij} + 4\pi \chi_{ij}(\omega, \mathbf{k}), \tag{2.8}$$

may be introduced in place of the dielectric permittivity tensor ϵ_{ij} for the characteristics of the electromagnetic properties of a medium. The susceptibility operator $\hat{\chi}_{ij}$ establishes the connection between the induced current and the time derivative of the field intensity

$$j_i = \hat{\chi}_{ij} \dot{E}_j . \tag{2.9}$$

This equation may be considered as the material equation supplementing the system (2.1).

2. The Wave Equation

Having solved the system of Eqs. (2.4) for the electrical field intensity $E_{\mathbf{k}\omega}$, we obtain

$$\Lambda_{ij}(\omega, \mathbf{k}) E_j(\mathbf{k}, \omega) = 0, \tag{2.10}$$

where

$$\Lambda_{ij}(\omega, \mathbf{k}) = \eta^2 \left(\frac{k_i k_j}{k^2} - \delta_{ij} \right) + \epsilon_{ij}(\omega, \mathbf{k}) \tag{2.11}$$

and $\eta = kc/\omega$ is the refractive index of the wave with frequency ω.

Equation (2.10) determines the electromagnetic field in a medium in the absence of external charges and currents, i.e., the electromagnetic waves in the medium. The compatibility condition for the separate components of the vector equation (2.10) reduces to requiring that the determinant composed of elements of the matrix Λ_{ij} vanish:

$$\Lambda \equiv \mid \Lambda_{ij}(\omega, \mathbf{k}) \mid = 0. \tag{2.12}$$

This equation is customarily called the dispersion equation. It implicitly determines the dispersion law, i.e., the dependence between the frequency and the wave vector for electromagnetic waves being propagated in the medium.

In the general case the determinant Λ is a complex function of ω and $\mathbf{k}$ hence, condition (2.12) reduces to the requirement that the real and imaginary parts of Λ vanish separately. In the plasma transparency domain (the domain of ω and $\mathbf{k}$ in which the anti-Hermitian part of the dielectric permittivity tensor is small as compared to the Hermitian part), the imaginary part of Λ is small compared to the real part of Λ. Hence, neglecting damping of the modes, the dispersion equation may be written approximately as

$$\text{Re } \Lambda(\omega, \mathbf{k}) = 0. \tag{2.13}$$

Finding the natural frequency of the mode ω_s from (2.13) and considering the damping to be small ($\gamma \ll \omega$) the damping coefficient of the wave may be found easily by using (2.12):

$$\gamma = \frac{\text{Im } \Lambda(\omega, \mathbf{k})}{(\partial/\partial\omega) \text{ Re } \Lambda(\omega, \mathbf{k})} \Big|_{\omega=\omega_s}. \tag{2.14}$$

Let us introduce the matrix λ_{ij} whose elements are the cofactors of the elements of the matrix Λ_{ij}. By definition

$$\Lambda_{ij}\lambda_{jk} = \Lambda \, \delta_{ik}. \tag{2.15}$$

The elements of the matrix λ_{ij} are expressed in terms of the elements of the matrix Λ_{ij} by means of the formula

$$\lambda_{ij} = \tfrac{1}{2}\epsilon_{ikl}\epsilon_{jmn}\Lambda_{mk}\Lambda_{nl},$$

where ϵ_{ikl} is a fully antisymmetric tensor. It is easy to see by direct substitution that Im $\varLambda$ may be expressed in terms of the Hermitian portion of the matrix λ_{ij} and the anti-Hermitian portion of the dielectric permittivity tensor ϵ_{ij} in the plasma transparency domain:

$$\text{Im } \varLambda = \frac{1}{4i} (\epsilon_{ij} - \epsilon_{ii}^*)(\lambda_{ji} + \lambda_{ij}^*). \tag{2.16}$$

Comparing (2.15) and (2.10), we see that for waves with the dispersion law (2.12) it is possible to select

$$e_j = C\lambda_{jk}a_k, \tag{2.17}$$

as the polarization vector, where a_k is an arbitrary unit vector and C is a constant determined from the normalization condition $ee^* = 1$.

Let us show that the relationship (2.17) defines the wave polarization vector to the accuracy of a phase constant. Let us first note that at any frequencies the matrices λ_{ij} and $\varLambda_{ij}$ are interrelated by means of the relation

$$\lambda_{ij}\lambda_{kl} = \lambda_{il}\lambda_{kj} + \varLambda\epsilon_{ikm}\epsilon_{jln}\varLambda_{nm} \tag{2.18}$$

(it is easy to see the validity of (2.18) by multiplying the left and right sides of the equality

$$\varLambda\epsilon_{abc} = \epsilon_{mnp}\varLambda_{ma}\varLambda_{nb}\varLambda_{pc}$$

by $\varLambda^{-1}\lambda_{aj}\lambda_{bl}\epsilon_{kic}$). At frequencies satisfying the dispersion law (2.12), the relationship (2.18) simplifies to

$$\lambda_{ij}\lambda_{kl} = \lambda_{il}\lambda_{kj}. \tag{2.19}$$

Neglecting the anti-Hermitian portion of λ_{ij} in the plasma transparency domain, the equality

$$\frac{\lambda_{il}a_l\lambda_{jk}^*a_k}{\lambda_{mn}a_m a_n} = \frac{\lambda_{il}a_l'\lambda_{jk}^*a_k'}{\lambda_{mn}a_m'a_n'} \tag{2.20}$$

is easily deduced from (2.19), where a and a' are arbitrary vectors. Let us also note that for arbitrary vectors a and a' the scalar products $(a\lambda a)$ and $(a'\lambda a')$ have the very same sign:

$$\lambda_{ij}a_i a_j \cdot \lambda_{kl}a_k'a_l' = | \lambda_{ij}a_i a_j' |^2 \tag{2.21}$$

(in particular, the diagonal elements of the Hermitian portion of the matrix λ_{ij} have the same sign).

Using (2.19), the normalization constant in (2.17) may be written as

$$C = (\lambda_{kl} a_k a_l \, \mathrm{Sp} \, \lambda)^{-1/2}.$$

Therefore, the normalized polarization vector for waves with the dispersion (2.12) is

$$e_i = \frac{\lambda_{ij} a_j}{(\lambda_{kl} a_k a_l \, \mathrm{Sp} \, \lambda)^{1/2}} \, . \tag{2.22}$$

According to (2.19), the product $e_i e_j{}^*$ is invariant relative to a change in the vector a; hence, only a phase factor can be changed in (2.22) for an arbitrary rotation in the vector a.

In the case of an isotropic medium characterized by the dielectric permittivity tensor (2.7), Eq. (2.10) splits into two separate equations for the longitudinal and transverse waves.

Longitudinal waves.   In the case of longitudinal waves $\mathbf{kE} \neq 0$, $[\delta_{ij} - (k_i k_j/k^2)]E_j = 0$ and (2.10) becomes

$$\epsilon_l(\omega, \mathbf{k}) \, \mathbf{E}(\mathbf{k}, \omega) = 0. \tag{2.23}$$

A nonzero solution of this equation is possible only if

$$\epsilon_l(\omega, \mathbf{k}) = 0. \tag{2.24}$$

The condition that the longitudinal dielectric permittivity ϵ_l vanish is indeed the dispersion equation for the longitudinal waves.

Transverse waves.   Now let us consider transverse waves in a medium for which $\mathbf{kE} = 0$ and $[\delta_{ij} - (k_i k_j/k^2)]E_j \neq 0$. In this case Eq. (2.10) becomes

$$\{\eta^2 - \epsilon_t(\omega, \mathbf{k})\} \, \mathbf{E}(\mathbf{k}, \omega) = 0. \tag{2.25}$$

The solution of this equation will be different from zero if the condition

$$\eta^2 - \epsilon_t(\omega, \mathbf{k}) = 0, \tag{2.26}$$

which is the dispersion equation for the transverse electromagnetic waves in the medium, is satisfied.

3. Dielectric and Magnetic Permittivities in a Medium with Spatial Dispersion

The dielectric and magnetic permittivities ϵ and μ may be introduced instead of the longitudinal and transverse dielectric permittivities ϵ_l and ϵ_t.

In the previous paragraphs we did not introduce the concept of the magnetic field intensity $\mathbf{H}$, and included the induced current in the medium $\mathbf{j}$ completely in the electric field induction $\mathbf{D}$. However, the induced current may be divided into polarized and vortical portions

$$\mathbf{j} = \dot{\mathbf{P}} + c \operatorname{rot} \mathbf{M}, \qquad \rho = -\operatorname{div} \mathbf{P}. \tag{2.27}$$

Then, defining the dielectric and magnetic permittivities by means of the equalities

$$\mathbf{D} = \mathbf{E} + 4\pi\mathbf{P} = \epsilon\mathbf{E},$$

$$\mathbf{H} = \mathbf{B} - 4\pi\mathbf{M} = \frac{1}{\mu}\mathbf{B}, \tag{2.28}$$

we transform the macroscopic system of Eqs. (2.1) to the customary form:

$$\operatorname{rot} \mathbf{E} = -\frac{1}{c}\frac{\partial \mathbf{B}}{\partial t}, \qquad \operatorname{div} \mathbf{B} = 0,$$

$$\operatorname{rot} \mathbf{H} = \frac{1}{c}\frac{\partial \mathbf{D}}{\partial t}, \qquad \operatorname{div} \mathbf{D} = 0. \tag{2.29}$$

It is easy to see that the dielectric and magnetic permittivities ϵ and μ thus defined are related to the longitudinal and transverse dielectric permittivities ϵ_l and ϵ_t by means of [Lindhard (3)]

$$\epsilon(\omega, \mathbf{k}) = \epsilon_l(\omega, \mathbf{k}),$$

$$\mu(\omega, \mathbf{k}) = \left\{1 + \frac{\omega^2}{k^2 c^2}[\epsilon_t(\omega, \mathbf{k}) - \epsilon_l(\omega, \mathbf{k})]\right\}^{-1}. \tag{2.30}$$

Henceforth, however, we shall use the quantities ϵ_l and ϵ_t.*

* The book by Silin and Rukhadze (4) is devoted to the analysis of the electromagnetic properties of media with spatial dispersion.

4. Current, Charge, and Field Fluctuations in a Medium

Let us now apply the general theory of fluctuations expounded in the previous section to the analysis of electromagnetic fluctuations in media with space-time dispersion.

In order to find the current-density fluctuations, let us assume that a random secondary field $\tilde{\mathbf{E}}$ has originated in the medium. This secondary field must additionally be included in the material equation which establishes the connection between the induced current and the field. Hence, we obtain in place of (2.9)

$$j_i = \frac{1}{4\pi}(\hat{\epsilon}_{ij} - \delta_{ij})(\dot{E}_j + \dot{\tilde{E}}_j). \tag{2.31}$$

The change in energy in the medium due to the effect of the secondary field is

$$\frac{\partial U}{\partial t} = \int d\mathbf{r}\, \tilde{\mathbf{E}}(\mathbf{r},\, t)\, \mathbf{j}(\mathbf{r},\, t). \tag{2.32}$$

Comparing this equation with (1.20), we see that the secondary field $\tilde{\mathbf{E}}$ should be considered as the quantity $-\dot{\mathbf{A}}$ in the analysis of the fluctuations in $\mathbf{j}$. The spectral distribution of the fluctuations of the current density $\mathbf{j}$ will hence be determined, according to (1.25), by the coefficient of proportionality between j_i and the quantity $A_j = (1/i\omega)E_j$. In order to find this coefficient it is necessary to use the system of Maxwell equations (2.1), in addition to the material equation (2.31). The system (2.1) may be written in Fourier components as

$$\Lambda^0_{ij} E_j(\mathbf{k},\, \omega) = \frac{4\pi}{i\omega}\, j_i(\mathbf{k},\, \omega), \tag{2.33}$$

where

$$\Lambda^0_{ij} \equiv \eta^2 \left(\frac{k_i k_j}{k^2} - \delta_{ij}\right) + \delta_{ij}. \tag{2.34}$$

Eliminating the current j_i from (2.33) with the aid of (2.31), we obtain the equation

$$\Lambda_{ij} E_j(\mathbf{k},\, \omega) = (\delta_{ij} - \epsilon_{ij})\, \tilde{E}_j(\mathbf{k},\, \omega),$$

whose solution may be written as

$$E_i(\mathbf{k}, \omega) = \Lambda_{ik}^{-1}(\delta_{kj} - \epsilon_{kj})\, \tilde{E}_j(\mathbf{k}, \omega). \tag{2.35}$$

(Λ_{ij}^{-1} is the inverse tensor to (2.11) and is determined from the condition $\Lambda_{ik}^{-1}\Lambda_{kj} = \delta_{ij}$.) Substituting (2.35) into (2.33) and noting that $\delta_{ij} - \epsilon_{ij} = \Lambda_{ij}^0 - \Lambda_{ij}$, we finally find

$$\alpha_{ij}(\omega, \mathbf{k}) = \frac{\omega^2}{4\pi}\{\Lambda_{ij}^0 - \Lambda_{ik}^0\Lambda_{kl}^{-1}\Lambda_{lj}^0\}. \tag{2.36}$$

According to (1.25), we obtain for the spectral distribution of the current-density fluctuations

$$\langle j_i j_j \rangle_{\mathbf{k}\omega} = \frac{i}{4\pi}\frac{\hbar\omega^2}{e^{\hbar\omega/T} - 1}\Lambda_{ik}^0\{\Lambda_{lk}^{-1} - \Lambda_{kl}^{-1*}\}\Lambda_{lj}^0. \tag{2.37}$$

The inverse tensor Λ_{ij}^{-1} is expressed in terms of the cofactor λ_{ij} and the determinant $\Lambda \equiv |\Lambda_{ij}|$ $(\lambda_{ik}\Lambda_{kj} = \Lambda\delta_{ij})$:

$$\Lambda_{ij}^{-1} = \frac{\lambda_{ij}}{\Lambda}. \tag{2.38}$$

The condition $\Lambda = 0$ is the dispersion equation for electromagnetic waves in the medium. Because of the presence of Λ in the denominator of the coefficient α_{ij} the spectral distribution of the fluctuations has sharp maximums near the frequencies of the intrinsic oscillations of the system. If wave damping in the medium is neglected, the spectral distribution of the fluctuations becomes

$$\langle j_i j_j \rangle_{\mathbf{k}\omega} = \frac{1}{2}\frac{\hbar\omega^2}{e^{\hbar\omega/T} - 1}\Lambda_{ik}^0\lambda_{kl}\Lambda_{lj}^0\delta\{\Lambda(\omega, \mathbf{k})\}, \tag{2.39}$$

i.e., the frequency spectrum of the fluctuations will contain only the natural frequencies of the field oscillations in the medium. Using the connection between the charge and current densities $\omega\rho_{\mathbf{k}\omega} = \mathbf{k}\mathbf{j}_{\mathbf{k}\omega}$, which results from the continuity equation (2.2), by using (2.37) it is easy to find the spectral distribution of the charge density fluctuations in the medium

$$\langle \rho^2 \rangle_{\mathbf{k}\omega} = \frac{1}{2\pi}\frac{\hbar}{e^{\hbar\omega/T} - 1}\,\mathrm{Im}\, k_i\Lambda_{ij}^{-1*}k_j. \tag{2.40}$$

Finally, by expressing the field $E_i(\mathbf{k}, \omega)$ in terms of $j_j(\mathbf{k}, \omega)$, and using (2.37), we find the spectral distribution of the electric field fluctuations in the medium

$$\langle E_i E_j \rangle_{\mathbf{k}\omega} = 4\pi i \, \frac{\hbar}{e^{\hbar\omega/T} - 1} \, \{A_{ji}^{-1} - A_{ij}^{-1*}\}. \qquad (2.41)$$

Let us note that both the fluctuations of the charge and current densities and the fluctuations of the electromagnetic field in the medium with space-time dispersion are determined completely by the dielectric permittivity tensor.

5. Fluctuations in an Isotropic Medium

In the case of an isotropic medium in the absence of constant external fields, the formulas for the spectral distributions of the fluctuations simplify in an essential manner.

Indeed, for the isotropic case both the dielectric permittivity tensor ϵ_{ij} and the tensor A_{ij} are separated into longitudinal and transverse parts:

$$A_{ij} = \frac{k_i k_j}{k^2} \epsilon_l + \left(\delta_{ij} - \frac{k_i k_j}{k^2}\right)(\epsilon_t - \eta^2). \qquad (2.42)$$

Hence, the longitudinal and transverse fluctuations in an isotropic medium turn out to be mutually independent. Separating the secondary field $\tilde{\mathbf{E}}$ into longitudinal and transverse parts $\tilde{\mathbf{E}}_l$ and $\tilde{\mathbf{E}}_t$, and using (2.35), it is easy to find the longitudinal and transverse components of the fluctuating field in the medium

$$\mathbf{E}_l(\mathbf{k}, \omega) = \frac{1 - \epsilon_l}{\epsilon_l} \tilde{\mathbf{E}}_l(\mathbf{k}, \omega), \qquad \mathbf{E}_t(\mathbf{k}, \omega) = \frac{1 - \epsilon_t}{\epsilon_t - \eta^2} \tilde{\mathbf{E}}_t(\mathbf{k}, \omega). \qquad (2.43)$$

The longitudinal and transverse induced currents in the medium $\mathbf{j}_l$ and $\mathbf{j}_t$ are hence connected to the longitudinal and transverse fields by means of the relationships

$$\mathbf{j}_l(\mathbf{k}, \omega) = \frac{i\omega}{4\pi} \mathbf{E}_l(\mathbf{k}, \omega), \qquad \mathbf{j}_t(\mathbf{k}, \omega) = \frac{i\omega}{4\pi} (1 - \eta^2) \mathbf{E}_t(\mathbf{k}, \omega). \qquad (2.44)$$

Substituting (2.43) into (2.44) and noting that $\tilde{\mathbf{E}}_{\mathbf{k}\omega} = i\omega\tilde{\mathbf{A}}_{\mathbf{k}\omega}$, we find for the isotropic case

$$j_i(\mathbf{k}, \omega) = \alpha_{ij}\tilde{A}_j(\mathbf{k}, \omega),$$

$$\alpha_{ij} = \frac{k_i k_j}{k^2}\alpha_l + \left(\delta_{ij} - \frac{k_i k_j}{k^2}\right)\alpha_t, \tag{2.45}$$

where α_l and α_t are the proportionality coefficients between the longitudinal and transverse components of the current $\mathbf{j}(\mathbf{k}, \omega)$ and the potential $\tilde{\mathbf{A}}(\mathbf{k}, \omega)$. These coefficients are expressed in terms of the longitudinal and transverse dielectric permittivities by means of the relations:

$$\alpha_l = \frac{\omega^2}{4\pi}\frac{\epsilon_l - 1}{\epsilon_l}, \qquad \alpha_t = \frac{\omega^2}{4\pi}(1 - \eta^2)\frac{\epsilon_t - 1}{\epsilon_t - \eta^2}. \tag{2.46}$$

The coefficients α_l and α_t are complex functions of the frequency ω and the modulus of the wave vector $\mathbf{k}$. The real parts of α_l and α_t are even functions of the frequency, and the imaginary parts are odd functions. The real and imaginary parts of the coefficients α_l and α_t are interrelated by means of the Kramers–Kronig relationships:

$$\alpha_l'(\omega, \mathbf{k}) - \alpha_l(\infty, \mathbf{k}) = \frac{1}{\pi}\int_{-\infty}^{\infty}\frac{\alpha_l''(\omega', \mathbf{k})}{\omega' - \omega}\,d\omega',$$

$$\alpha_t'(\omega, \mathbf{k}) - \alpha_t(\infty, \mathbf{k}) = \frac{1}{\pi}\int_{-\infty}^{\infty}\frac{\alpha_t''(\omega', \mathbf{k})}{\omega' - \omega}\,d\omega'. \tag{2.47}$$

The spectral distribution of the current density fluctuations in an isotropic medium is determined by the formula

$$\langle j_i j_j\rangle_{\mathbf{k}\omega} = \frac{2\hbar}{e^{\hbar\omega/T} - 1}\left\{\frac{k_i k_j}{k^2}\,\mathrm{Im}\,\alpha_l + \left(\delta_{ij} - \frac{k_i k_j}{k^2}\right)\mathrm{Im}\,\alpha_t\right\} \tag{2.48}$$

or if the relationships (2.36) are used, by the formula

$$\langle j_i j_j\rangle_{\mathbf{k}\omega} = \frac{1}{2\pi}\frac{\hbar\omega^2}{e^{\hbar\omega/T} - 1}\left\{\frac{k_i k_j}{k^2}\frac{\mathrm{Im}\,\epsilon_l}{|\epsilon_l|^2} + \left(\delta_{ij} - \frac{k_i k_j}{k^2}\right)(1 - \eta^2)^2\frac{\mathrm{Im}\,\epsilon_t}{|\epsilon_t - \eta^2|^2}\right\}. \tag{2.49}$$

The first member in (2.49) describes the longitudinal current fluctuations, while the second member describes the transverse fluctuations.

Because of the continuity equation (2.2) the charge density fluctuations are connected to the longitudinal current fluctuations. The spectral distribution of the charge density fluctuations in the isotropic case is determined by the formula

$$\langle \rho^2 \rangle_{\mathbf{k}\omega} = \frac{k^2}{2\pi} \frac{\hbar}{e^{\hbar\omega/T} - 1} \frac{\text{Im } \epsilon_l}{|\epsilon_l|^2}. \tag{2.50}$$

Using (2.44), we obtain the following formula for the spectral distribution of the electric field intensity fluctuations from (2.49):

$$\langle E_i E_j \rangle_{\mathbf{k}\omega} = 8\pi \frac{\hbar}{e^{\hbar\omega/T} - 1} \left\{ \frac{k_i k_j}{k^2} \frac{\text{Im } \epsilon_l}{|\epsilon_l|^2} + \left(\delta_{ij} - \frac{k_i k_j}{k^2} \right) \frac{\text{Im } \epsilon_t}{|\epsilon_t - \eta^2|^2} \right\}. \tag{2.51}$$

Finally, using the connection between the magnetic and electric fields $\mathbf{B}_{\mathbf{k}\omega} = (c/\omega)\mathbf{k} \times \mathbf{E}_{\mathbf{k}\omega}$, we find for the magnetic field fluctuations

$$\langle B_i B_j \rangle_{\mathbf{k}\omega} = 8\pi \frac{\hbar}{e^{\hbar\omega/T} - 1} \left(\delta_{ij} - \frac{k_i k_j}{k^2} \right) \eta^2 \frac{\text{Im } \epsilon_t}{|\epsilon_t - \eta^2|^2}. \tag{2.52}$$

For sufficiently high temperatures $T \gg \hbar\omega$ it is possible to pass to the classical limit in formulas (2.49)–(2.52) by replacing the factor $\hbar/(e^{\hbar\omega/T} - 1)$ by (T/ω). In the classical case the current density and charge density fluctuations are determined by the formulas:

$$\langle j_i j_j \rangle_{\mathbf{k}\omega} = \frac{\omega}{2\pi} T \left\{ \frac{k_i k_j}{k^2} \frac{\text{Im } \epsilon_l}{|\epsilon_l|^2} + \left(\delta_{ij} - \frac{k_i k_j}{k^2} \right) (1 - \eta^2)^2 \frac{\text{Im } \epsilon_t}{|\epsilon_t - \eta^2|^2} \right\}, \tag{2.53}$$

$$\langle \rho^2 \rangle_{\mathbf{k}\omega} = \frac{k^2}{2\pi} \frac{T}{\omega} \frac{\text{Im } \epsilon_l}{|\epsilon_l|^2}. \tag{2.54}$$

If the inequality $T \gg \hbar\omega$ is valid for all frequencies for which the spectral distributions of the fluctuations differ from zero in an essential manner, then by using the Kramers–Kronig integral relations these distributions may be integrated with respect to the frequency in the general case. Indeed, by writing the spectral distribution of the current density fluctuations in the form

$$\langle j_i j_j \rangle_{\mathbf{k}\omega} = 2 \frac{T}{\omega} \left\{ \frac{k_i k_j}{k^2} \text{Im } \alpha_l + \left(\delta_{ij} - \frac{k_i k_j}{k^2} \right) \text{Im } \alpha_t \right\}$$

and taking account of (2.47), we find at once

$$\langle j_i j_j \rangle_{\mathbf{k}} = T \left\{ \frac{k_i k_j}{k^2} \left[\alpha_l(0, \mathbf{k}) - \alpha_l(\infty, \mathbf{k}) \right] \right.$$

$$\left. + \left(\delta_{ij} - \frac{k_i k_j}{k^2} \right) \left[\alpha_t(0, \mathbf{k}) - \alpha_t(\infty, \mathbf{k}) \right] \right\}, \qquad (2.55)$$

where $\alpha_l(\omega, \mathbf{k})$ and $\alpha_t(\omega, \mathbf{k})$ are determined by the relationships (2.46). We have here taken into account that the imaginary parts of α_l and α_t vanish as $\omega \to 0$ and $\omega \to \infty$. In an analogous manner, we have for the charge density fluctuations

$$\langle \rho^2 \rangle_{\mathbf{k}} = \frac{k^2}{4\pi} T \left\{ 1 - \frac{1}{\epsilon_l(0, \mathbf{k})} \right\}. \qquad (2.56)$$

By using the integral relations (2.47) it is also possible to find the mean square frequency of the fluctuations in general form. For example, we determine the mean square frequency of the charge density fluctuations $\langle \omega^2 \rangle$ by means of the relationship

$$\langle \omega^2 \rangle = \int_{-\infty}^{\infty} \omega^2 \langle \rho^2 \rangle_{\mathbf{k}\omega} \, d\omega \Big/ \int_{-\infty}^{\infty} \langle \rho^2 \rangle_{\mathbf{k}\omega} \, d\omega. \qquad (2.57)$$

Using (2.37), it is easy to see that this frequency equals

$$\langle \omega^2 \rangle = \frac{\Omega^2}{1 - \epsilon_l(0, \mathbf{k})^{-1}}, \qquad (2.58)$$

where $\Omega^2 = \lim_{\omega \to \infty} \omega^2 \{ 1 - \epsilon_l(\omega, \mathbf{k}) \}$. Since for sufficiently high frequencies the spatial dispersion is negligible $\epsilon_l(\omega, \mathbf{k}) \to \epsilon(\omega)$, the asymptotic frequency dependence of the customary dielectric permittivity

$$\epsilon(\omega) = 1 - \frac{4\pi e^2 n_0}{m\omega^2}, \qquad \omega \to \infty$$

may be used to find Ω^2 (n_0 is the equilibrium electron density, e and m the electron charge and mass). Therefore

$$\Omega^2 = \frac{4\pi e^2 n_0}{m}. \qquad (2.59)$$

In exactly the same manner the mean square frequency of the transverse fluctuations may be found.

6. Inversion of the Fluctuation–Dissipation Theorem

According to the general theory expounded in the previous sections, the electromagnetic fluctuations in a medium, particularly the charge and current density fluctuations, are determined completely by the dielectric permittivity tensor characterizing the macroscopic properties of the substance.

Conversely, if the charge and current density fluctuations are determined directly by using microscopic theory, the fluctuation–dissipation theorem may be used to find the tensor of the dielectric permittivity of the substance.* The characteristic peculiarity of such a determination of ϵ_{ij} is the possibility of taking account of thermal effects without utilization of the kinetic equation.

In the case of an isotropic medium the electromagnetic properties are described completely by the longitudinal and transverse dielectric permittivities ϵ_l and ϵ_t. According to (2.50) and (2.49), the spectral distributions of the correlation functions of the charge and transverse-current density fluctuations are expressed in terms of ϵ_l and ϵ_t by utilizing the relations:

$$\langle \rho^2 \rangle_{\mathbf{k}\omega} = \frac{k^2}{2\pi} \frac{\hbar}{e^{\hbar\omega/T} - 1} \frac{\operatorname{Im} \epsilon_l}{|\epsilon_l|^2},$$

$$\langle j_t^2 \rangle_{\mathbf{k}\omega} = 2 \frac{\omega^2}{2\pi} \frac{\hbar}{e^{\hbar\omega/T} - 1} (1 - \eta^2)^2 \frac{\operatorname{Im} \epsilon_t}{|\epsilon_t - \eta^2|^2}. \tag{2.60}$$

Let us expand the left- and right-hand sides of (2.60) in power series in e^2, and let us retain the basic terms. Then remarking that the longitudinal and transverse electrical susceptibilities are proportional to e^2, we obtain

$$\langle \rho \rangle_{\mathbf{k}\omega}^0 = \frac{k^2}{2\pi} \frac{\hbar}{e^{\hbar\omega/T} - 1} \operatorname{Im} \epsilon_l, \qquad \langle j_t^2 \rangle_{\mathbf{k}\omega}^0 = 2 \frac{\omega^2}{2\pi} \frac{\hbar}{e^{\hbar\omega/T} - 1} \operatorname{Im} \epsilon_t, \tag{2.61}$$

where $\langle \rho^2 \rangle_{\mathbf{k}\omega}^0$ and $\langle j_t^2 \rangle_{\mathbf{k}\omega}^0$ are the spectral distributions of the correlation functions for systems of noninteracting particles. Inverting the first of the equalities (2.61) we have

$$\operatorname{Im} \epsilon_l(\omega, \mathbf{k}) = \frac{2\pi}{\hbar k^2} (e^{\hbar\omega/T} - 1) \langle \rho^2 \rangle_{\mathbf{k}\omega}^0, \tag{2.62}$$

---

* Shafranov (5), Kubo (6), and Nakano (7) proposed such a method of determining the macroscopic coefficients of a substance.

i.e., the imaginary part of the longitudinal dielectric permittivity is expressed directly in terms of the spectral distribution of the charged-particle density fluctuations without taking account of the Coulomb interaction. The real part of the dielectric permittivity ϵ_l is easily found by utilizing the first of the relations (2.47):

$$\operatorname{Re} \epsilon_l(\omega, \mathbf{k}) = 1 + \frac{2}{\hbar k^2} \int_{-\infty}^{\infty} (e^{\hbar\omega'/T} - 1) \frac{\langle \rho^2 \rangle_{\mathbf{k}\omega'}^0}{\omega' - \omega} \, d\omega'. \qquad (2.63)$$

Formulas (2.62) and (2.63) may be combined by representing the longitudinal dielectric permittivity in the form

$$\epsilon_l(\omega, \mathbf{k}) = 1 + \frac{2}{\hbar k^2} \int_{-\infty}^{\infty} (e^{\hbar\omega'/T} - 1) \frac{\langle \rho^2 \rangle_{\mathbf{k}\omega'}^0}{\omega' - \omega - io} \, d\omega'. \qquad (2.64)$$

The presence of the member io in the denominator of the integrand means that the integration with respect to ω' is along a contour passing along the real axis in the complex ω' plane with the singular point $\omega' = \omega$ bypassed from below.

In an analogous manner, on the basis of the second equality of (2.61) and the Kramers–Kronig relationship for α_t, we have

$$\epsilon_t(\omega, \mathbf{k}) = 1 - \frac{\Omega^2}{\omega^2} + \frac{1}{\hbar\omega^2} \int_{-\infty}^{\infty} (e^{\hbar\omega'/T} - 1) \frac{\langle j_t^2 \rangle_{\mathbf{k}\omega'}^0}{\omega' - \omega - io} \, d\omega'. \qquad (2.65)$$

Formulas (2.64) and (2.65) determine the dielectric permittivities of a substance $\epsilon_l(\omega, k)$ and $\epsilon_t(\omega, k)$, by means of the spectral distributions of the charge and transverse-current density fluctuations $\langle \rho^2 \rangle_{\mathbf{k}\omega}^0$ and $\langle j_t^2 \rangle_{\mathbf{k}\omega}^0$, computed without taking account of the interaction between the charged particles.

In the general case, in the absence of isotropy, the electromagnetic properties of a substance are characterized by the tensor ϵ_{ij}. To find this tensor, let us start from the general expression (2.37). It is easy to see by direct substitution that the expansion

$$\Lambda_{kl}^{-1} = (\Lambda_{kl}^0)^{-1} - (\Lambda_{km}^0)^{-1}(\epsilon_{mn} - \delta_{mn})(\Lambda_{nl}^0)^{-1} + \cdots$$

is valid to the accuracy of terms on the order of e^2. Expanding the right- and left-hand sides of (2.37) in power series in e^2, and keeping the basic terms, we therefore find

$$\langle j_i j_j \rangle_{\mathbf{k}\omega}^0 = \frac{i}{4\pi} \frac{\hbar\omega^2}{e^{\hbar\omega/T} - 1} \{\epsilon_{ij}^* - \epsilon_{ji}\}, \qquad (2.66)$$

where $\langle j_i j_j \rangle^0_{\mathbf{k}\omega}$ is the spectral distribution of the current fluctuations in the medium without taking account of the electromagnetic interaction between the particles.

Let us separate the symmetric and antisymmetric parts out of the dielectric permittivity tensor ϵ_{ij}

$$\epsilon_{ij} = \epsilon^s_{ij} + \epsilon^a_{ij}, \qquad \epsilon^s_{ij} = \epsilon^s_{ji}, \qquad \epsilon^a_{ij} = - \epsilon^a_{ji}. \tag{2.67}$$

The energy dissipation in the medium is determined by the imaginary part of ϵ^s_{ij} and the real part of ϵ^a_{ij}. These dissipative parts of the dielectric permittivity tensor are determined from (2.66). Thus, Im ϵ^s_{ij} is expressed in terms of the symmetrized correlation function, and Re ϵ^a_{ij} in terms of the antisymmetrized correlation function. The parts of the tensor ϵ_{ij}, not associated with the dissipation of the field energy in the medium are easily found by utilizing the integral relations (1.38) and (1.39). We finally obtain the following general formula determining the dielectric permittivity tensor of a substance $\epsilon_{ij}(\omega, \mathbf{k})$ by means of the known spectral distribution of the current fluctuations in the medium $\langle j_i j_j \rangle^0_{\mathbf{k}\omega}$:

$$\epsilon_{ij}(\omega, \mathbf{k}) = \left(1 - \frac{\Omega^2}{\omega^2}\right)\delta_{ij} + \frac{2}{\hbar\omega^2} \int_{-\infty}^{\infty} (e^{\hbar\omega'/T} - 1) \frac{\langle j_j j_i \rangle^0_{\mathbf{k}\omega'}}{\omega' - \omega - io} \, d\omega'. \tag{2.68}$$

In the classical case $(T \gg \hbar\omega')$ formula (2.68) simplifies to

$$\epsilon_{ij}(\omega, \mathbf{k}) = \delta_{ij} + \frac{2}{\omega T} \int_{-\infty}^{\infty} \frac{\langle j_j j_i \rangle^0_{\mathbf{k}\omega}}{\omega' - \omega - io} \, d\omega'. \tag{2.69}$$

[We used the relationship $\int_{-\infty}^{\infty} \langle j_i j_j \rangle^0_{\mathbf{k}\omega'} \, d\omega' = \frac{1}{2} T\Omega^2 \delta_{ij}$ in going from (2.68) to (2.69).] The components of the dielectric permittivity tensor of a substance (2.68) are analytic functions in the complex ω plane. Comparing (2.68) with (1.48) it is easy to see that the components of $\epsilon_{ij}(\omega, \mathbf{k})$ are expressed directly in terms of the spectral representation of the retarding Green's function for the currents $G^R_{ij}(\mathbf{k}, \omega)$.

The dielectric permittivity tensor of a material under non-equilibrium conditions may be determined analogously by using the fluctuation-dissipation relationship in the form (1.23)

$$\epsilon_{ij}(\omega, \mathbf{k}) = \left(1 - \frac{\Omega^2}{\omega^2}\right)\delta_{ij} + \frac{2}{\hbar\omega^2} \int_{-\infty}^{\infty} \frac{(\langle j_j j_i \rangle^{\hbar\omega'}_{\mathbf{k}\omega'})^0 - \langle j_j j_i \rangle^0_{\mathbf{k}\omega'}}{\omega' - \omega - io} \, d\omega'. \tag{2.70}$$

In contrast to the equilibrium case (2.68), in order to determine the dielectric permittivity tensor $\epsilon_{ij}(\omega, \mathbf{k})$ in a nonequilibrium system, it is necessary to give the spectral distributions of the two correlation functions for the free currents $\langle j_i j_j \rangle^0_{\mathbf{k}\omega}$ and $(\langle j_i j_j \rangle^{\hbar\omega}_{\mathbf{k}\omega})^0$. In the limiting classical case ($\hbar \to 0$) the expression for the dielectric permittivity tensor of a material under nonequilibrium conditions may be represented as

$$\epsilon_{ij}(\omega, \mathbf{k}) = \left(1 - \frac{\Omega^2}{\omega^2}\right)\delta_{ij} - \frac{2}{\omega^2}\int_{-\infty}^{\infty} \frac{\omega'(\partial/\partial E)\langle j_j j_i \rangle^0_{\mathbf{k}\omega'}}{\omega' - \omega - io}\, d\omega'. \tag{2.71}$$

REFERENCES

1. M. E. Gertsenshtein, *ZETF* **22**, 303 (1952).
2. A. G. Sitenko and K. N. Stepanov, *ZETF* **31**, 642 (1956) *(Sov. Phys. JETP)*.
3. J. Lindhard, *Dan. Mat. Fys. Medd.* **28**(8) (1954).
4. V. P. Silin and A. A. Rukhadze, "Electromagnetic Properties of Plasmas and Plasma-Like Media." Atomizdat, Moscow, 1961 (English translation, Gordon & Breach, New York, 1965).
5. V. D. Shafranov, "Plasma Physics and the Problem of Controlled Thermonuclear Reactions," Vol. 4, p. 416. AN SSSR Press, Moscow, 1961. (English translation, Pergamon Press, New York).
6. R. Kubo, *J. Phys. Soc. Japan* **12**, 570 (1957).
7. H. Nakano, *Progr. Theor. Phys.* **15**, 77 (1954); **17**, 145 (1957).

3
Electrodynamic Properties
of an Electron Plasma

1. Space-Time Dispersion in Plasma

In comparison with the properties of other material media, the electrodynamic properties of a plasma are characterized by a number of peculiarities. In particular, when considering alternating electromagnetic fields in a plasma, it is necessary to take into account not only the time but also the space dispersion.

The presence of space-time dispersion means that the induced charges and currents in the plasma at a specific time depend on the values of the fields at all preceding times, and that a nonlocal connection holds between the currents and the fields. Hence, the material coefficients (dielectric permittivity tensor) characterizing the properties of the plasma, turn out to depend on both the frequency and on the wave vector of the alternating electromagnetic field.

Neglecting the spatial dispersion in customary media is based on the smallness of the atomic dimensions as compared with the wavelength of the electromagnetic fields. In the case of plasma, the characteristic space parameter is the Debye radius, whose magnitude is commensurate with the wavelength; hence, spatial dispersion substantially affects the plasma properties. Thus, the spatial dispersion turns out to be connected to such specific plasma properties as the shielding of electric charge in the plasma, the existence of charge density oscillations in the plasma, etc.

The plasma dielectric permittivity tensor may be determined by finding the spectral distributions of the charge and current density fluctuations without taking into account the interaction between the charged particles in the plasma, and by utilizing the inversion of the fluctuation–dissipation theorem [Shafranov (1)].

2. Space-Time Correlation Functions for a System of Noninteracting Particles

Let us consider an unbounded, homogeneous, fully ionized plasma. The ion motion may be neglected because of their large mass. In this case the presence of ions in the plasma will be manifest only in cancellation of the equilibrium electric charge. We shall designate a plasma where the ion motion is neglected completely, as an electron plasma.

Let us determine the electron density at a point with radius–vector $\mathbf{r}$ at time t by means of the relationship

$$n(\mathbf{r}, t) = \sum_{\alpha} \delta(\mathbf{r} - \mathbf{r}_{\alpha}(t)), \tag{3.1}$$

where $\mathbf{r}_{\alpha}(t)$ is the radius–vector of some electron α at time t, and the summation is over all electrons in a unit volume. Neglecting interaction between electrons, the radius–vector $\mathbf{r}_{\alpha}(t)$ may be represented as

$$\mathbf{r}_{\alpha}(t) = \mathbf{r}_{\alpha} + \mathbf{v}_{\alpha}t, \tag{3.2}$$

where $\mathbf{r}_{\alpha}$ and $\mathbf{v}_{\alpha}$ are the radius–vector and velocity of the electron at the initial instant $t = 0$.

Let n_0 denote the mean value of the electron density:

$$n_0 \equiv \langle n(\mathbf{r}, t) \rangle. \tag{3.3}$$

The brackets $\langle \cdots \rangle$ denote the statistical averaging operation; in particular, the averaging operation is made for the thermodynamic equilibrium state over the equilibrium electron distribution.

$$\langle \cdots \rangle = \int \prod_{\alpha} d\tau_{\alpha} \, e^{-(E_{\alpha}/T)} \cdots \Big/ \int \prod_{\alpha} d\tau_{\alpha} \, e^{-(E_{\alpha}/T)}, \tag{3.4}$$

where $d\tau_{\alpha} = d\mathbf{r}_{\alpha} \, d\mathbf{v}_{\alpha}$ is an element of the phase volume and

$E_\alpha = mv_\alpha^2/2$ (m is the electron mass and T the temperature in energy units). The electron-density fluctuations are characterized by the deviation of the density n from the equilibrium value n_0

$$\delta n(\mathbf{r}, t) = n(\mathbf{r}, t) - n_0. \qquad (3.5)$$

Let us determine the correlation function of the electron-density fluctuations at different points of space ($\mathbf{r}_1$ and $\mathbf{r}_2$) at different times (t_1 and t_2).

$$\langle \delta n^2 \rangle_{rt}^0 \equiv \langle \delta n(\mathbf{r}_1, t_1)\, \delta n(\mathbf{r}_2, t_2) \rangle, \qquad (3.6)$$

which will depend only on the difference in the point coordinates $\mathbf{r} = \mathbf{r}_2 - \mathbf{r}_1$ and the difference in the times $t = t_2 - t_1$, because of the homogeneity of the space and the isotropy of the time. Utilizing (3.1) and (3.5) we find

$$\langle \delta n^2 \rangle_{rt}^0 = \left\langle \sum_\alpha \delta(\mathbf{r}_1 - \mathbf{r}_\alpha - \mathbf{v}_\alpha t_1)\, \delta(\mathbf{r}_2 - \mathbf{r}_\alpha - \mathbf{v}_\alpha t_2) \right\rangle. \qquad (3.7)$$

If we introduce the one-particle distribution function $f_0(v)$, normalized according to the condition $\int dv\, f_0(v) = 1$, the correlation function (3.7) may then be represented in the form

$$\langle \delta n^2 \rangle_{rt}^0 = n_0 \int dv\, f_0(v)\, \delta(\mathbf{r} - \mathbf{v}t). \qquad (3.8)$$

Selecting the Maxwell distribution function $f_0(v) = (m/2\pi T)^{3/2}$ $\exp(-mv^2/2T)$ as the equilibrium distribution function, we obtain the following expression for the correlation function

$$\langle \delta n^2 \rangle_{rt}^0 = n_0 \left(\frac{m}{2\pi T} \right)^{3/2} \frac{\exp\left[-(m/2T)(r^2/t^2) \right]}{|t|^3}. \qquad (3.9)$$

In the relativistic case

$$f_0(v) = A\, \frac{c^2}{(c^2 - v^2)^{5/2}} \exp\left[-(mc^2/T)\, \frac{1}{(1 - (v^2/c^2))^{1/2}} \right],$$
$$A^{-1} = 4\pi\, \frac{T}{mc^2}\, K_2\left(\frac{mc^2}{T} \right) \qquad (3.10)$$

[$K_2(x)$ the MacDonald function], and we find for the correlation function of the density fluctuations

$$\langle \delta n^2 \rangle^0_{\mathbf{r}t} = A n_0 \frac{c^2 t^2}{(c^2 t^2 - r^2)^{5/2}} \exp\left[-(mc^2/T)\left(\frac{c^2 t^2}{c^2 t^2 - r^2}\right)^{1/2}\right]. \quad (3.11)$$

Because of the hypothesized homogeneity of the ion background, the correlation function for the charge-density fluctuations in an electron plasma is directly connected with (3.8) by means of the relation

$$\langle \rho^2 \rangle^0_{\mathbf{r}t} = e^2 \langle \delta n^2 \rangle^0_{\mathbf{r}t}. \quad (3.12)$$

Let us determine the correlation function for the current-density fluctuations. The density of the current associated with the electron motion is determined by the expression

$$\mathbf{j}(\mathbf{r}, t) = \sum_\alpha e v_\alpha \delta(\mathbf{r} - \mathbf{r}_\alpha(t)). \quad (3.13)$$

Let us note that the mean value of the current equals zero; hence, the current fluctuations are determined directly from (3.13). Utilizing (3.13), the correlation function for the current fluctuations is obtained as

$$\langle j_i j_j \rangle^0_{\mathbf{r}t} = e^2 n_0 \int d\mathbf{v}\, v_i v_j\, f_0(v)\, \delta(\mathbf{r} - \mathbf{v}t). \quad (3.14)$$

It is easy to verify that the correlation functions for the current and charge density fluctuations are connected by means of the relationship

$$\langle j_i j_j \rangle^0_{\mathbf{r}t} = \frac{x_i x_j}{t^2} \langle \rho^2 \rangle^0_{\mathbf{r}t}. \quad (3.15)$$

In conclusion, let us present expressions for the spectral distributions of the correlation functions (3.12) and (3.14):

$$\langle \rho^2 \rangle^0_{\mathbf{k}\omega} = 2\pi e^2 n_0 \int d\mathbf{v}\, f_0(v)\, \delta(\omega - \mathbf{k}\mathbf{v}), \quad (3.16)$$

$$\langle j_i j_j \rangle^0_{\mathbf{k}\omega} = 2\pi e^2 n_0 \int d\mathbf{v}\, v_i v_j\, f_0(v)\, \delta(\omega - \mathbf{k}\mathbf{v}). \quad (3.17)$$

For example, we have in the case of the Maxwell equilibrium function

$$\langle \rho^2 \rangle^0_{\mathbf{k}\omega} = \sqrt{6\pi}\, \frac{e^2 n_0}{ks} \exp -\left(\frac{3}{2} \frac{\omega^2}{k^2 s^2}\right) \quad (3.18)$$

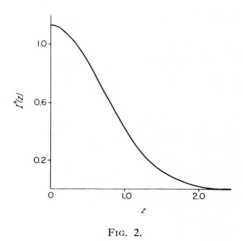

FIG. 2.

(s is the mean-square electron velocity). Integration of (3.18) with respect to the frequencies yields

$$\langle \rho^2 \rangle_{\mathbf{k}}^0 = e^2 n_0 . \tag{3.19}$$

The spectral distribution of the density fluctuations of non-interacting particles $I^0(z) = 2\langle \rho^2 \rangle_{\mathbf{k}z}/\langle \rho^2 \rangle_{\mathbf{k}}$ is represented in Fig. 2 as a function of the nondimensional frequency $z = \sqrt{\frac{3}{2}}\,(\omega/ks)$.

3. Determination of the Plasma Dielectric Permittivity Tensor

Knowing the spectral distribution of the current fluctuations of the free electrons (3.17), the plasma dielectric permittivity tensor may easily be evaluated on the basis of inverting the fluctuation-dissipation relationship (2.71). According to the classical formula (3.17), the quantity $\hbar\omega$ characterizes the change in energy associated with the motion of an individual particle along the vector $\mathbf{k}$, hence

$$\omega \frac{\partial}{\partial E} \langle j_i j_j \rangle_{\mathbf{k}\omega}^0 = \frac{2\pi e^2 n_0}{m} \int v_i v_j \mathbf{k} \frac{\partial f_0}{\partial \mathbf{v}} \delta(\omega - \mathbf{k}\mathbf{v})\, dv,$$

Substituting this expression into (2.71) we obtain the following general formula for the plasma dielectric permittivity tensor:

$$\epsilon_{ij}(\omega, \mathbf{k}) = \left(1 - \frac{\Omega^2}{\omega^2}\right)\delta_{ij} - \frac{\Omega^2}{\omega^2}\int \frac{v_i v_j \mathbf{k}\, \partial f_0/\partial \mathbf{v}}{\mathbf{kv} - \omega - io}\, d\mathbf{v}. \qquad (3.20)$$

The presence of the member $-io$ in the denominator of the integrand in (3.20) means that the integration with respect to the longitudinal velocity component v_l should be along a contour passing along the real axis in the complex v_l plane and bypassing the singular point $v_l = \omega/\mathbf{k}$ from below.

In the isotropic case, i.e., in the absence of external constant electric and magnetic fields, the plasma dielectric permittivity tensor ϵ_{ij} is characterized just by two independent components, the longitudinal and transverse components of the dielectric permittivity ϵ_l and ϵ_t :

$$\epsilon_{ij}(\omega, \mathbf{k}) = \frac{k_i k_j}{k^2}\, \epsilon_l(\omega, \mathbf{k}) + \left(\delta_{ij} - \frac{k_i k_j}{k^2}\right)\epsilon_t(\omega, \mathbf{k}).$$

These components may be found directly by means of known spectral distributions of the charge and transverse-current density fluctuations $\langle\rho^2\rangle^0_{\mathbf{k}\omega}$ and $\langle j_t^2\rangle^0_{\mathbf{k}\omega}$, evaluated without taking account of interaction between electrons.

Using (3.16) and (3.17) for the spectral distributions $\langle\rho^2\rangle^0_{\mathbf{k}\omega}$ and $\langle j_t^2\rangle^0_{\mathbf{k}\omega}$, we obtain the plasma dielectric permittivities in the classical case as

$$\epsilon_l(\omega, \mathbf{k}) = 1 - \frac{\Omega^2}{k^2}\int \frac{\mathbf{k}(\partial f_0(v)/\partial \mathbf{v})}{\mathbf{kv} - \omega - io}\, d\mathbf{v}, \qquad (3.21)$$

$$\epsilon_t(\omega, \mathbf{k}) = 1 - \frac{1}{2}\frac{\Omega^2}{\omega k^2}\int \frac{[[\mathbf{kv}]\mathbf{v}](\partial f_0(v)/\partial \mathbf{v})}{\mathbf{kv} - \omega - io}\, d\mathbf{v}. \qquad (3.22)$$

Selecting the Maxwell distribution as the equilibrium function, and integrating with respect to the velocity components perpendicular to the vector $\mathbf{k}$ in (3.21) and (3.22), we find

$$\epsilon_l(\omega, \mathbf{k}) = 1 + \frac{1}{a^2 k^2}\{1 - \varphi(z) + i\sqrt{\pi}z e^{-z^2}\}, \qquad (3.23)$$

$$\epsilon_t(\omega, \mathbf{k}) = 1 - \frac{\Omega^2}{\omega^2}\{\varphi(z) - i\sqrt{\pi}z e^{-z^2}\}, \qquad (3.24)$$

where $z = \sqrt{\frac{3}{2}}\,(\omega/ks)$, $(a^2 = T/4\pi e^2 n_0$ is the square of the Debye radius), and $\varphi(z)$ is the real function

$$\varphi(z) \equiv \frac{z}{\sqrt{\pi}} \int_{-\infty}^{\infty} \frac{e^{-y^2}}{z-y}\, dy. \tag{3.25}$$

Using the Fok transformation (2), the function $\varphi(z)$ may be transformed to

$$\varphi(z) = 2z\, e^{-z^2} \int_{0}^{z} e^{x^2}\, dx. \tag{3.26}$$

Let us present the asymptotic expansion for $\varphi(z)$ for small and large values of the argument

$$\varphi(z) = 2z^2 + \cdots , \qquad\qquad z \ll 1, \tag{3.27}$$

$$\varphi(z) = 1 + \frac{1}{2z^2} + \frac{3}{4z^4} + \cdots , \quad z \gg 1. \tag{3.28}$$

The dependence of the function $\varphi(z)$ on z is presented in Fig. 3.

According to (3.23) and (3.24), the dielectric permittivities ϵ_l and ϵ_t are complex, which indicates the presence of energy dissipation in the plasma even in the absence of collisions. It is easy to understand the dissipation mechanism if attention is turned to

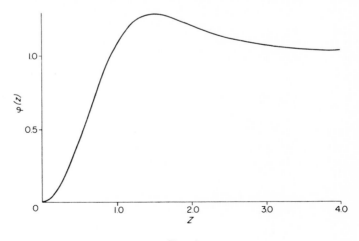

FIG. 3.

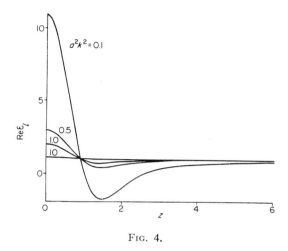

FIG. 4.

the following circumstance. The imaginary parts in ϵ_l and ϵ_t are specified by the residue at the point $v_l = \omega/k$ during integration with respect to the particle velocities. Therefore, particles whose velocities satisfy the condition $\mathbf{kv} = \omega$, which is the condition for both Cerenkov radiation of particles and for the inverse Cerenkov effect by which radiation is absorbed by the particles, will yield a contribution to the imaginary parts of the dielectric permittivities.

Presented in Figs. 4–7 are the real and imaginary parts of the dielectric permittivities ϵ_l and ϵ_t as a function of the nondimensional

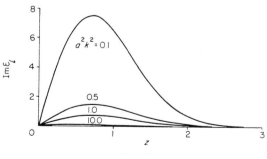

FIG. 5.

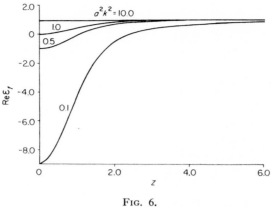

frequency z for different values of the parameter a^2k^2. We also present values of the dielectric permittivities ϵ_l and ϵ_t for $k \to 0$ and $\omega \to 0$. In the limiting case of long waves ($k \to 0$) the spatial dispersion is negligible, and the dielectric permittivities ϵ_l and ϵ_t turn out to be equal

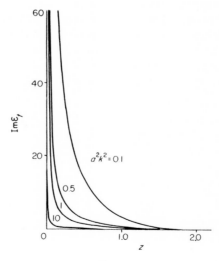

FIG. 7.

$$\epsilon_l(\omega, 0) = \epsilon_t(\omega, 0) = 1 - \frac{\Omega^2}{\omega^2}. \tag{3.29}$$

In the limiting case of low frequencies ($\omega \to 0$) we obtain the statical value different from one for the longitudinal dielectric permittivity ϵ_l:

$$\epsilon_l(0, \mathbf{k}) = 1 + \frac{1}{a^2 k^2}. \tag{3.30}$$

In the low-frequency range the transverse dielectric permittivity ϵ_t may be represented as

$$\epsilon_t(\omega, \mathbf{k}) = 1 + \frac{4\pi i \sigma}{\omega}, \tag{3.31}$$

where $\sigma = (1/4 \sqrt{2\pi})(\Omega/ak)$ is the transverse static conductivity of the plasma.

4. Electromagnetic Waves in Plasma

The electrodynamic properties of a plasma are determined completely by the dielectric permittivity tensor $\epsilon_{ij}(\omega, \mathbf{k})$; in particular, electromagnetic waves being propagated in a plasma are determined completely by the dielectric permittivity tensor. In the isotropic plasma case, characterized by the tensor (3.20), the wave equation decomposes into two separate equations for the longitudinal and transverse waves.

Longitudinal waves. The dispersion equation governing the connection between the frequency ω and the wave vector $\mathbf{k}$ reduces, for longitudinal waves in a plasma, to the condition that the longitudinal dielectric permittivity should vanish*

$$\epsilon_l(\omega, \mathbf{k}) = 0.$$

Utilizing (3.23) for the dielectric permittivity ϵ_l, this dispersion equation may be written as

$$1 + \frac{1}{a^2 k^2} \{1 - \varphi(z) + i \sqrt{\pi} z\, e^{-z^2}\} = 0, \qquad z = \sqrt{\frac{3}{2}} \frac{\omega}{ks}. \tag{3.32}$$

---

* Langmuir (3) first considered longitudinal oscillations in plasma.

In the case of weak spatial dispersion ($ak \ll 1$) it is not difficult to find an approximate solution of (3.32) by using the expansion (3.28) for $\varphi(z)$. It turns out that for a given real value of k the corresponding frequency is complex, i.e., longitudinal waves are damped in a plasma. The squares of the frequency and the damping coefficient of longitudinal waves are determined by the formulas of Vlasov (4) and Landau (5):

$$\omega^2 = \Omega^2 + k^2 s^2, \tag{3.33}$$

$$\gamma = \sqrt{\frac{\pi}{8}} \frac{\Omega}{a^3 k^3} \exp - \left[\frac{1}{2a^2 k^2} + \frac{3}{2} \right]. \tag{3.34}$$

Damping of the longitudinal waves grows strongly as the wavelength diminishes. If $k \sim a^{-1}$, the phase velocity of the longitudinal wave is commensurate with the thermal electron velocity. Hence, the number of electrons for which the condition of the inverse Cerenkov effect $\omega = \mathbf{kv}$ is satisfied, is large, which indeed leads to strong damping of the oscillations.

Oscillations of the longitudinal field in a plasma are accompanied by oscillations in the charge density ($i\mathbf{kE_{k\omega}} = 4\pi\rho_{\mathbf{k}\omega}$); hence, the longitudinal waves are customarily designated plasma waves. Let us note that plasma oscillations are a collective effect, due to the Coulomb interaction between electrons.

Transverse waves. The dispersion equation for transverse electromagnetic waves in a plasma reduces to the requirement

$$\epsilon_t(\omega, \mathbf{k}) - \frac{k^2 c^2}{\omega^2} = 0.$$

By using (3.24) this equation may be rewritten thus

$$1 - \frac{\Omega^2}{\omega^2} \{\varphi(z) - i \sqrt{\pi} z \, e^{-z^2}\} - \frac{k^2 c^2}{\omega^2} = 0, \qquad z = \sqrt{\frac{3}{2}} \frac{\omega}{ks}. \tag{3.35}$$

If the thermal electron motion is neglected completely, we find for the square of the transverse-wave frequency from (3.35)

$$\omega^2 = \Omega^2 + k^2 c^2. \tag{3.36}$$

The phase velocity of the transverse waves hence turns out to be greater than the velocity of light ($\omega/k = c/[1 - (\Omega^2/\omega^2)]^{1/2} > c$). Hence, for a nonrelativistic plasma ($s \ll c$), the corrections to (3.36) associated with taking account of the thermal electron motion are very small

$$\omega^2 = \left(1 + \frac{1}{3}\frac{k^2 s^2}{\omega^2}\right)\Omega^2 + k^2 c^2. \tag{3.37}$$

It is impossible to use the nonrelativistic formula (3.35) to find the damping of the transverse waves. Actually, the imaginary part of the dielectric permittivity ϵ_t is due to particles whose velocities coincide with the phase velocity of the waves. Since the phase velocities of transverse waves in plasma exceed the velocity of light, the imaginary part of ϵ_t should vanish (a rigorous relativistic analysis arrives at the same result), and therefore, there is no damping of transverse waves in a plasma.

The smallness of the thermal corrections in (3.37) and the absence of damping mean that spatial dispersion may be neglected completely in analyzing transverse waves in plasma, and the approximate expression (3.29) may be used as the transverse dielectric permittivity.

5. Relativistic Plasma

Let us consider the electrodynamic properties of a relativistic electron plasma. Evidently relativistic effects will play an essential part in plasmas if the thermal electron energy T is comparable to their rest energy.

As before, the spectral distributions of the correlation functions for a relativistic plasma will be determined by (3.16) and (3.17), in which, however, (3.10) must be used as the equilibrium distribution. Integrating with respect to the velocities, we find

$$\langle\rho^2\rangle_{\mathbf{k}\omega}^0 = \frac{\pi e^2 n_0}{kc} K_2\left(\frac{mc^2}{T}\right)^{-1}\left\{\frac{1}{1 - (\omega^2/k^2c^2)} + 2\frac{T}{mc^2}\frac{1}{[1 - (\omega^2/k^2c^2)]^{1/2}}\right.$$
$$\left. + 2\left(\frac{T}{mc^2}\right)^2\right\}\exp\left[-\frac{mc^2}{T}\frac{1}{[1 - (\omega^2/k^2c^2)]^{1/2}}\right]\theta(kc - |\omega|),$$

$$\tag{3.38}$$

$$\langle j_t^2 \rangle_{\mathbf{k}\omega}^0 = \frac{2\pi e^2 n_0 c}{k} \frac{T}{mc^2} K_2 \left(\frac{mc^2}{T} \right)^{-1} \left\{ \left(1 - \frac{\omega^2}{k^2 c^2} \right)^{1/2} \right.$$

$$\left. + \frac{T}{mc^2} \left(1 - \frac{\omega^2}{k^2 c^2} \right) \right\} \exp \left[- \frac{mc^2}{T} \frac{1}{[1 - (\omega^2/k^2 c^2)]^{1/2}} \right] \theta(kc - |\omega|). \tag{3.39}$$

Here $\theta(x)$ is the Heaviside function (1.42). Substituting the spectral distributions $\langle \rho^2 \rangle_{\mathbf{k}\omega}^0$ and $\langle \mathbf{j}_t^2 \rangle_{\mathbf{k}\omega}^0$ into (2.64) and (2.65), we obtain the following expressions [Silin (6)] for the dielectric permittivities ϵ_l and ϵ_t of a relativistic plasma:

$$\epsilon_l(\omega, \mathbf{k}) = 1 + \frac{\Omega^2}{k^3 c^3} K_2 \left(\frac{mc^2}{T} \right)^{-1} \int_{-kc}^{kc} \exp \left\{ - \frac{mc^2}{T} \frac{1}{[1 - (\omega'^2/k^2 c^2)]^{1/2}} \right\}$$

$$\times \left[\frac{T}{mc^2} + \frac{1}{[1 - (\omega'^2/k^2 c^2)]^{1/2}} + \frac{mc^2}{T} \frac{1}{[1 - (\omega'^2/k^2 c^2)]} \right]$$

$$\times \frac{\omega'}{\omega' - \omega - io} d\omega', \tag{3.40}$$

$$\epsilon_t(\omega, \mathbf{k}) = 1 + \frac{\Omega^2}{2\omega kc} K_2 \left(\frac{mc^2}{T} \right)^{-1} \int_{-kc}^{kc} \exp \left\{ - \frac{mc^2}{T} \frac{1}{[1 - (\omega'^2/k^2 c^2)]^{1/2}} \right\}$$

$$\times \left[[1 - (\omega'^2/k^2 c^2)]^{1/2} + \frac{T}{mc^2} \left(1 - \frac{\omega'^2}{k^2 c^2} \right) \right] \frac{1}{\omega' - \omega - io} d\omega'. \tag{3.41}$$

In the limiting case of long waves $(k \to 0)$ the longitudinal and transverse dielectric permittivities agree and are determined by the expression

$$\epsilon_l(\omega, 0) = \epsilon_t(\omega, 0) = 1 - \frac{\tilde{\Omega}^2}{\omega^2}, \tag{3.42}$$

$$\tilde{\Omega}^2 = \Omega^2 K_2 \left(\frac{mc^2}{T} \right)^{-1} \int_0^1 \exp \left[- \frac{mc^2}{T} \frac{1}{(1 - \xi^2)^{1/2}} \right]$$

$$\times \left[(1 - \xi^2)^{1/2} + \frac{T}{mc^2} (1 - \xi^2) \right] d\xi. \tag{3.43}$$

For an ultrarelativistic electron plasma $(T \gg mc^2)$ the longitudinal and transverse dielectric permittivities are determined for ω and $k > 0$ by the formulas:

$$\epsilon_l(\omega, \mathbf{k}) = 1 + \frac{1}{a^2 k^2}\left\{1 + \frac{\omega}{2kc}\ln\frac{|\omega - kc|}{\omega + kc} + i\frac{\pi}{4}\frac{\omega}{kc}\left(1 - \frac{\omega - kc}{|\omega - kc|}\right)\right\},$$
$$(3.44)$$

$$\epsilon_t(\omega, \mathbf{k}) = 1 - \frac{1}{2a^2 k^2}\left\{1 - \frac{kc}{2\omega}\left(1 - \frac{\omega^2}{k^2 c^2}\right)\ln\frac{|\omega - kc|}{\omega + kc} - i\frac{\pi}{4}\frac{kc}{\omega}\right.$$

$$\left. \times \left(1 - \frac{\omega^2}{k^2 c^2}\right)\left(1 - \frac{\omega - kc}{|\omega - kc|}\right)\right\}.$$
$$(3.45)$$

By using these formulas the following expressions may be obtained for the proper frequencies of the longitudinal and transverse oscillations of an ultrarelativistic plasma:

$$\omega^2 = \frac{c^2}{3a^2} + \frac{3}{5}k^2 c^2, \quad \omega \gg kc; \qquad \omega = (1 + 2\,e^{-2(1+a^2 k^2)})kc, \quad \omega \to kc;$$
$$(3.46)$$

$$\omega^2 = \frac{c^2}{3a^2} + \frac{6}{5}k^2 c^2, \quad \omega \gg kc; \qquad \omega^2 = \frac{c^2}{2a^2} + k^2 c^2, \quad \omega \to kc.$$
$$(3.47)$$

According to (3.44) and (3.45), the imaginary parts of ϵ_l and ϵ_t vanish if $\omega > kc$. Since the phase velocities of the longitudinal and transverse waves exceed the velocity of light, there is no wave damping in an ultrarelativistic plasma.

6. Nonequilibrium Plasma and Stability

Electromagnetic waves in an equilibrium electron plasma were considered in the preceding sections. A characteristic peculiarity of waves in such a plasma is the presence of damping even when binary collisions are neglected. The damping is connected with the resonant interaction between the particles and waves and is due to the fact that under equilibrium conditions the particle distribution function decreases monotonely as the energy grows; hence, particles moving in phase with the wave receive more energy from the wave than they emit.

A sufficiently rarefied and hot plasma may be in the nonequilibrium state during considerable time intervals. The properties of waves in such a plasma may differ substantially from the equilibrium case. If a nonequilibrium particle distribution in a plasma does not decrease monotonely with the energy, then the energy transferred by the wave to the particles will not absolutely exceed the energy received by the wave from the particles; hence, undamped oscillations are possible in the plasma and the appropriate particle distribution turns out to be unstable.

Using (3.21), which is also valid in the nonequilibrium case, for the longitudinal dielectric permittivity of the plasma, it is easy to see that the imaginary part of the frequency of the longitudinal high-frequency oscillations is determined directly by the derivative of the distribution function for values of the velocity equal to the phase velocity of the wave

$$\text{Im } \omega \equiv -\gamma = \frac{\pi}{2} \frac{\Omega^2}{k} v \frac{\partial f_0(v)}{\partial v}\Big|_{v=\text{Re }\omega/k}, \qquad (3.48)$$

where $f_0(v_\parallel) = \int f_0(\mathbf{v}) \, d\mathbf{v}_\perp$. If the distribution function decreases monotonely as the energy of the particles grows, then $\partial f_0/\partial v < 0$ and the oscillations are damped ($\gamma > 0$). If the distribution is such that $\partial f_0/\partial v > 0$ in a specific range of velocities, then the oscillations with the phase velocities turn out to be increasing ($\gamma < 0$) in the mentioned interval.

The simplest example of a nonequilibrium plasma is a plasma through which a compensated electron beam passes. Let the electron densities of the plasma and the beam be denoted by n_0 and n_0', and let us assume that the plasma and beam electrons are characterized by the Maxwell distributions f_0 and f_0' with the temperatures T and T':

$$f_0 = \left(\frac{m}{2\pi T}\right)^{3/2} \exp\left(-\frac{mv^2}{2T}\right), \quad f_0' = \left(\frac{m}{2\pi T'}\right)^{3/2} \exp\left(-\frac{m(\mathbf{v} - \mathbf{u})^2}{2T'}\right),$$

$$\qquad (3.49)$$

where $\mathbf{u}$ is the mean velocity of directed motion of the beam electrons.

If the velocity $\mathbf{u}$ is small compared with the velocity of light c, then the longitudinal and transverse components may be separated out in the dielectric permittivity tensor (3.20); the longitudinal and transverse waves hence turn out to be independent. Considering the condition $u \ll c$ to be satisfied, we obtain for the longitudinal dielectric permittivity

$$\epsilon_l(\omega, \mathbf{k}) = 1 + \frac{1}{a^2 k^2} \{1 - \varphi(z) + i \sqrt{\pi} \, z e^{-z^2}\}$$

$$+ \frac{1}{a'^2 k^2} \{1 - \varphi(y) + i \sqrt{\pi} \, y e^{-y^2}\}, \tag{3.50}$$

where

$$y = \sqrt{\frac{3}{2}} \frac{\omega - \mathbf{ku}}{ks'}, \qquad s'^2 = \frac{3T'}{m}, \qquad \text{and} \qquad a'^2 = \frac{T'}{4\pi e^2 n_0'}.$$

We shall seek the solution of the dispersion equation for the longitudinal waves

$$\epsilon_l(\omega, \mathbf{k}) = 0$$

as a power series in n_0'/n_0. If the consideration is limited to a low density beam $n_0' \ll n_0$, then the contribution of the beam to the real part of the plasma dielectric permittivity may be neglected. The dispersion law for the weakly damped longitudinal waves will hence be exactly the same as for the free plasma. The contribution of the beam to the imaginary part of the dielectric permittivity, which determines the damping decrement, turns out to be substantial. Using (3.50) we obtain the following expression for the decrement of the oscillations:

$$\frac{\gamma}{\omega} = \sqrt{\frac{\pi}{8}} \frac{1}{a^3 k^3} \frac{\omega}{\Omega} \left\{ \exp\left(-\frac{3}{2} \frac{\omega^2}{k^2 s^2}\right) + \frac{n_0'}{n_0} \left(\frac{T}{T'}\right)^{3/2} \frac{\omega - \mathbf{ku}}{\omega} \right.$$

$$\left. \times \exp\left(-\frac{3}{2} \frac{(\omega - \mathbf{ku})^2}{k^2 s'^2}\right) \right\}. \tag{3.51}$$

It is easy to see that for a sufficiently small beam velocity $\gamma > 0$, that is, the oscillations are damped. At some critical velocity $\tilde{u}$, the decrement γ vanishes and becomes negative as the beam velocity increases further. Hence, for $u > \tilde{u}$ the longitudinal oscillations with wave vector $\mathbf{k}$ do not damp, but grow; the corresponding state of the plasma-beam system hence turns out to be unstable (7, 8).

In the case of a sufficiently hot beam $(\omega - \mathbf{ku})^2 \ll k^2 s'^2$ the critical velocity equals

$$\tilde{u} = \frac{\omega}{k} \left\{ 1 + \frac{n_0}{n_0'} \left(\frac{T'}{T}\right)^{3/2} \exp\left(-\frac{3}{2} \frac{\omega^2}{k^2 s^2}\right) \right\}. \tag{3.52}$$

Substituting $\omega = (\Omega^2 + k^2 s^2)^{1/2}$ into (3.52), it is easily seen that the instability of the system originates at a beam velocity u on the order of the thermal velocity of the plasma electrons s.

As another example of a nonequilibrium system let us consider a plasma with an anisotropic velocity distribution. In this case the instability originates because of the growing nature of the transverse perturbations in the system.

Let us characterize the plasma by a Maxwell distribution with temperature anisotropies $T_\perp$ and $T_\parallel$:

$$f_0 = \frac{m}{2\pi T_\perp}\left(\frac{m}{2\pi T_\parallel}\right)^{1/2}\exp\left(-\frac{mv_\perp{}^2}{2T_\perp} - \frac{mv_\parallel{}^2}{2T_\parallel}\right). \tag{3.53}$$

Longitudinal and transverse parts:

$$\epsilon_l(\omega, \mathbf{k}) = 1 + \frac{\Omega^2}{k^2 s_\parallel{}^2}\{1 - \varphi(z) + i\sqrt{\pi}\,z\,e^{-z^2}\}, \tag{3.54}$$

$$\epsilon_t(\omega, \mathbf{k}) = 1 - \frac{\Omega^2}{\omega^2}\left\{1 - \frac{T_\perp}{T_\parallel}[1 - \varphi(z) + i\sqrt{\pi}\,z\,e^{-z^2}]\right\}, \tag{3.55}$$

where $z = \omega/\sqrt{2}\,ks_\parallel$ and $s_\parallel{}^2 = T_\parallel/m$, may be separated out in the dielectric permittivity tensor (3.20) for waves being propagated along the axis of system symmetry.

The dispersion equation for the transverse waves in the plasma is

$$\frac{k^2 c^2}{\omega^2} = \epsilon_t(\omega, \mathbf{k}).$$

Considering the phase velocity of the waves to be considerably less than the thermal velocity of the electrons ($z \ll 1$), this equation may be written as

$$k^2 = \frac{\Omega^2}{c^2}\left\{\frac{T_\perp}{T_\parallel} - 1 + i\sqrt{\pi}\,\frac{T_\perp}{T_\parallel}z\right\}, \tag{3.56}$$

from which follows

$$\omega = i\left(\frac{2}{\pi}\right)^{1/2}\frac{T_\parallel}{T_\perp}\frac{(k_0{}^2 - k^2)c^2}{\Omega^2}ks_\parallel, \tag{3.57}$$

where

$$k_0{}^2 = \left(\frac{T_\perp}{T_\parallel} - 1\right)\frac{\Omega^2}{c^2}.$$

If $T_\perp < T_\parallel$, then $k_0{}^2 < 0$ and the appropriate perturbations are damped. If $T_\perp > T_\parallel$, then $k_0{}^2 > 0$. In this case perturbations with $k < k_0$ are increasing (Im $\omega > 0$), and, therefore, the state of the system is unstable.

Hence, the state of a plasma with anisotropic temperature distribution turns out to be unstable relative to transverse perturbations with the wave vectors $k < k_0$ (9) for $T_\perp > T_\parallel$.

REFERENCES

1. V. D. Shafranov, "Electromagnetic Waves in Plasmas." IAE AN SSSR, Moscow, 1960.
2. V. A. Fok, "Radiowave Diffraction around the Earth's Surface." AN SSSR Press, Moscow, 1946 (English translation available from MDF).
3. I. Langmuir, *Proc. Natl. Acad. Sci.* 14, 627 (1926).
4. A. A. Vlasov, *ZETF* 8, 291 (1938).
5. L. D. Landau, *ZETF* 16, 574 (1946) *(J. Phys. USSR)*.
6. V. P. Silin, *ZETF* 38, 1577 (1960); 40, 616 (1961) *(Sov. Phys.—JETP)*.
7. A. I. Akhiezer and Ia. B. Fainberg, *DAN SSSR* 69, 555 (1949); *ZETF* 21, 1262 (1951).
8. D. Bohm and E. P. Gross, *Phys. Rev.* 75, 1851 (1949).
9. A. E. Stefanovich, *ZTF* 32, 638 (1962) *(Sov. Phys.—Tech. Phys.)*

4

Electromagnetic Fluctuations
in an Electron Plasma

---

1. Fluctuations in an Electron Plasma

Now let us turn to the analysis of fluctuations in an electron plasma taking account of the interaction between charged particles. Since we neglect ion motion, the charge and current densities are determined completely by the electron density and motion. Hence, fluctuations in such a plasma are of purely electromagnetic nature, i.e., are determined completely by the dielectric permittivity tensor.

Using the expressions found in the previous chapter for the dielectric permittivities of an electron plasma ϵ_l and ϵ_t, we may, on the basis of the fluctuation–dissipation formulas, find the spectral distributions for both the fluctuations of the charge and current densities in the plasma, and for the fluctuations of the electromagnetic fields. We will limit ourselves herein to the analysis of classical fluctuations in a plasma ($T \gg \hbar\omega$).

Taking account of the interaction between electrons leads to an essential change in the nature of the spectral distributions of the fluctuations, especially in the high-frequency range of the spectra where collective effects are manifested most strongly. In particular, additional sharp maximums, corresponding to the proper oscillations of the plasma (1), are manifest in the spectra of the charge- and current-density fluctuations in a plasma, because of the interaction between electrons.

2. Charge Density Fluctuations

First, let us consider the charge-density fluctuations in an electron plasma. According to (2.54) and (2.55), the charge-density fluctuations are determined in the classical case by the formulas

$$\langle \rho^2 \rangle_{\mathbf{k}\omega} = \frac{k^2}{2\pi} \frac{T}{\omega} \frac{\mathrm{Im}\, \epsilon_l}{|\epsilon_l|^2}, \qquad \langle \rho^2 \rangle_{\mathbf{k}} = \frac{k^2}{4\pi} T \left\{ 1 - \frac{1}{\epsilon_l(0, \mathbf{k})} \right\}. \tag{4.1}$$

Utilizing the expression (3.23) for the plasma dielectric permittivity ϵ_l, we obtain from the second relationship of (4.1)

$$\langle \rho^2 \rangle_{\mathbf{k}} = \frac{e^2 n_0 k^2}{k^2 + (4\pi e^2 n_0/T)}. \tag{4.2}$$

The spatial correlation function for the charge-density fluctuations is hence determined by the expression

$$\langle \rho^2 \rangle_{\mathbf{r}} = \frac{1}{(2\pi)^3} \int d\mathbf{k} \frac{e^2 n_0 k^2}{k^2 + (1/a^2)} e^{i\mathbf{k}\mathbf{r}}.$$

Integrating with respect to $d\mathbf{k}$, we find

$$\langle \rho^2 \rangle_{\mathbf{r}} = e^2 n_0 \left\{ \delta(\mathbf{r}) - \frac{1}{4\pi a^2} \frac{e^{-(r/a)}}{r} \right\}. \tag{4.3}$$

Therefore, the correlation between charge-density fluctuations in a plasma holds at ranges on the order of the Debye radius.

The static value of the plasma dielectric permittivity $\epsilon_l(0, \mathbf{k})$ also determines the mean-square frequency of the fluctuations, according to (2.58). Using (2.58) and (3.23), we find for the mean-square frequency of the charge-density fluctuations,

$$\langle \omega^2 \rangle = \Omega^2 + \tfrac{1}{3} k^2 s^2. \tag{4.4}$$

Let us note that the mean-square frequency of the fluctuations $\langle \omega^2 \rangle^{1/2}$ is less than the frequency of the proper plasma oscillations $\omega_p = (\Omega^2 + k^2 s^2)^{1/2}$.

According to (4.1) and (3.23), the spectral distribution of the charge-density fluctuations in a plasma is determined by the following general formula:

$$\langle \rho^2 \rangle_{\mathbf{k}\omega} = \sqrt{2\pi}\, e^2 n_0 \frac{a^3 k^3}{\Omega} \frac{e^{-z^2}}{[a^2 k^2 + 1 - \varphi(z)]^2 + \pi z^2 e^{-2z^2}}, \qquad z = \sqrt{\frac{3}{2}} \frac{\omega}{ks}. \tag{4.5}$$

In the low-frequency case this formula yields

$$\langle \rho^2 \rangle_{\mathbf{k}\omega} = \langle \rho^2 \rangle_{\mathbf{k}0} \frac{\tilde{\omega}^2}{\tilde{\omega}^2 + \omega^2}, \qquad \omega \ll ks, \tag{4.6}$$

where

$$\langle \rho^2 \rangle_{\mathbf{k}0} = \sqrt{2\pi} \, \frac{e^2 n_0}{\Omega} \frac{a^3 k^3}{(1 + a^2 k^2)^2}, \qquad \tilde{\omega}^2 = \frac{2}{\pi} a^2 k^2 (1 + a^2 k^2) \Omega^2.$$

In the high-frequency case, the approximate expression

$$\langle \rho^2 \rangle_{\mathbf{k}\omega} = \frac{2\pi}{3} e^2 n_0 \frac{k^2 s^2}{\omega} \delta(\omega^2 - \Omega^2 - k^2 s^2), \qquad \omega \gg ks \tag{4.7}$$

may be obtained from (4.5). Therefore, in the high-frequency range of the spectrum charge-density fluctuations in a plasma are possible only with frequencies close to the proper frequency of the plasma oscillations. The integral contribution of the high-frequency fluctuations to the total intensity (4.2) is found by integrating (4.7) with respect to the frequencies

$$\langle \tilde{\rho}^2 \rangle_{\mathbf{k}} = \frac{e^2 n_0 k^2}{3k^2 + (4\pi e^2 n_0 / T)}. \tag{4.8}$$

The ratio of the high-frequency to the total intensity is

$$\langle \tilde{\rho}^2 \rangle_{\mathbf{k}} / \langle \rho^2 \rangle_{\mathbf{k}} = \frac{1 + a^2 k^2}{1 + 3a^2 k^2}. \tag{4.9}$$

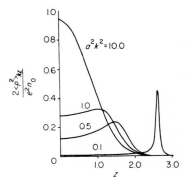

FIG. 8.

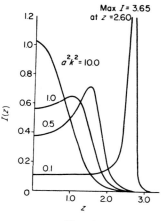

FIG. 9.

Presented in Figs. 8 and 9 is the spectral distribution of the charge-density fluctuations as a function of the nondimensional frequency $z = \sqrt{\tfrac{3}{2}}(\omega/ks)$ for various values of the parameter a^2k^2. Values of the function $I(z) = 2\langle \rho^2 \rangle_{kz}/\langle \rho^2 \rangle_k$ are plotted along the vertical axis in Fig. 9. (The function $I(z)$ satisfies the normalization condition $\int_0^\infty I(z)\,dz = 1$.) We see that the low-frequency fluctuations play a fundamental part for large values of the parameter a^2k^2. As the parameter a^2k^2 decreases the effective fluctuation frequency increases. For $a^2k^2 \ll 1$ only frequencies close to the proper frequency of the density oscillations in the plasma remain in the fluctuation spectrum.

Let us present the spectral distribution of the charge-density fluctuations for the ultrarelativistic plasma. Using (3.44), we find according to (4.1):

$$\langle \rho^2 \rangle_{k\omega} = \frac{\pi e^2 n_0}{\omega} \left\{ \frac{\omega}{kc} \frac{\theta(kc - \omega)}{\left[1 + \dfrac{1}{a^2k^2}\left(1 + \dfrac{\omega}{2kc}\ln\dfrac{|\omega - kc|}{\omega + kc}\right)\right]^2 + \dfrac{\pi^2}{4}\dfrac{\omega^2}{k^2c^2}\dfrac{1}{a^4k^4}} \right.$$

$$\left. + \tfrac{1}{2}a^2k^2\,\delta\left[1 + \frac{1}{a^2k^2}\left(1 + \frac{\omega}{2kc}\ln\frac{|\omega - kc|}{\omega + kc}\right)\right] \right\}. \tag{4.10}$$

Let us note that integration of (4.10) with respect to the frequencies leads to formula (4.2), exactly as in the nonrelativistic case.

3. Current-Density Fluctuations

Now let us consider the current–density fluctuations in an electron plasma. According to (2.53), the current fluctuations in the classical case are determined by the formula

$$\langle j_i j_j \rangle_{\mathbf{k}\omega} = \frac{\omega}{2\pi} T \left\{ \frac{k_i k_j}{k^2} \frac{\operatorname{Im} \epsilon_l}{|\epsilon_l|^2} + \left(\delta_{ij} - \frac{k_i k_j}{k^2} \right) (1 - \eta^2)^2 \frac{\operatorname{Im} \epsilon_t}{|\epsilon_t - \eta^2|^2} \right\}.$$

(4.11)

Utilizing (3.23) and (3.24) for the dielectric permittivities ϵ_l and ϵ_t of an electron plasma, we obtain the following formula for the spectral distribution of the current–density fluctuations:

$$\langle j_i j_j \rangle_{\mathbf{k}\omega} = \frac{\omega}{2\pi} T \left\{ \frac{k_i k_j}{k^2} \frac{\sqrt{\pi} z\, e^{-z^2}}{[a^2 k^2 + 1 - \varphi(z)]^2 + \pi z^2\, e^{-2z^2}} \right.$$

$$+ \left(\delta_{ij} - \frac{k_i k_j}{k^2} \right) \frac{\Omega^2}{\omega^2} (1 - \eta^2)^2$$

$$\left. \times \frac{\sqrt{\pi} z\, e^{-z^2}}{\left[\eta^2 - 1 + \dfrac{\Omega^2}{\omega^2} \varphi(z) \right]^2 + \pi \dfrac{\Omega^4}{\omega^4} z^2\, e^{-2z^2}} \right\}.$$

(4.12)

The first member in (4.12) characterizes the longitudinal current fluctuations, and the second, the transverse–current fluctuations. Let us present approximate expressions for the spectral distribution of the transverse-current fluctuations in the low and high-frequency ranges. If $\omega \ll kc$, then

$$\langle \mathbf{j}_t^2 \rangle_{\mathbf{k}\omega} = \sqrt{\frac{3}{2\pi}}\, T\, \frac{\Omega^2}{ks} \exp\left[-\frac{3}{2} \frac{\omega^2}{k^2 s^2} \right].$$

(4.13)

For $\omega \gg ks\ (\omega \sim kc)$ we have

$$\langle \mathbf{j}_t^2 \rangle_{\mathbf{k}\omega} = T\, \frac{\Omega^4}{\omega}\, \delta(\omega^2 - \Omega^2 - k^2 c^2),$$

(4.14)

i.e., only frequencies close to the proper frequencies of the transverse plasma oscillations are contained in the high-frequency range of the fluctuation spectrum.

By integrating the spectral distribution (4.12) with respect to the frequencies, we may find the Fourier component of the spatial correlation function of the current-density fluctuations in a plasma $\langle j_i j_j \rangle_\mathbf{k}$.

Remarking that $\alpha_l(0, \mathbf{k}) = \alpha_t(0, \mathbf{k}) = 0$ and $\alpha_l(\infty, \mathbf{k}) = \alpha_t(\infty, \mathbf{k}) = -(\Omega^2/4\pi)$ for an electron plasma, we find according to (2.55)

$$\langle j_i j_j \rangle_\mathbf{k} = \tfrac{1}{3} e^2 n_0 s^2 \delta_{ij} . \tag{4.15}$$

Using (4.15) it is easy to see that

$$\langle j_i j_j \rangle_\mathbf{r} = \tfrac{1}{3} e^2 n_0 s^2 \delta(\mathbf{r}) \, \delta_{ij} , \tag{4.16}$$

i.e., that a spatial correlation between the current–density fluctuations in an electron plasma is absent.

4. Electromagnetic Field Fluctuations

The spectral distributions of the electric and magnetic field fluctuations in a plasma are determined by the general formulas (2.51) and (2.52)

$$\langle E_i E_j \rangle_{\mathbf{k}\omega} = 8\pi \, \frac{T}{\omega} \left\{ \frac{k_i k_j}{k^2} \, \frac{\operatorname{Im} \epsilon_l}{|\epsilon_l|^2} + \left(\delta_{ij} - \frac{k_i k_j}{k^2} \right) \frac{\operatorname{Im} \epsilon_t}{|\epsilon_t - \eta^2|^2} \right\},$$

$$\langle B_i B_j \rangle_{\mathbf{k}\omega} = 8\pi \, \frac{T}{\omega} \left(\delta_{ij} - \frac{k_i k_j}{k^2} \right) \eta^2 \, \frac{\operatorname{Im} \epsilon_t}{|\epsilon_t - \eta^2|^2} . \tag{4.17}$$

Using (3.23) and (3.24) for the dielectric permittivities, we obtain the spectral distributions of the field fluctuations as

$$\langle E^2 \rangle_{\mathbf{k}\omega} = 8\pi \, \frac{T}{\omega} \, a^2 k^2 \left\{ \frac{\sqrt{\pi} z \, e^{-z^2}}{[a^2 k^2 + 1 - \varphi(z)]^2 + \pi z^2 \, e^{-2z^2}} \right.$$

$$+ 2 \, \frac{\Omega^2}{\omega^2} \, \frac{\sqrt{\pi} z \, e^{-z^2}}{\left[\eta^2 - 1 + \dfrac{\Omega^2}{\omega^2} \, \varphi(z) \right]^2 + \pi \, \dfrac{\Omega^4}{\omega^4} \, z^2 \, e^{-2z^2}} \left. \right\}, \tag{4.18}$$

$$\langle B^2 \rangle_{\mathbf{k}\omega} = 16\pi \, \frac{T}{\omega} \, \frac{\Omega^2}{\omega^2} \, \eta^2 \, \frac{\sqrt{\pi} z \, e^{-z^2}}{\left[\eta^2 - 1 + \dfrac{\Omega^2}{\omega^2} \, \varphi(z) \right]^2 + \pi \, \dfrac{\Omega^4}{\omega^4} \, z^2 \, e^{-z^2}} . \tag{4.19}$$

Integration of (4.18) and (4.19) with respect to the frequencies yields

$$\langle E^2 \rangle_{\mathbf{k}} = 4\pi T \left\{ \frac{1}{1 + a^2 k^2} + 2 \right\}, \qquad \langle B^2 \rangle_{\mathbf{k}} = 8\pi T. \qquad (4.20)$$

With the aid of (4.20) we easily find the spatial correlation functions for the electric and magnetic fields in a plasma

$$\langle E^2 \rangle_{\mathbf{r}} = 8\pi T \left\{ \delta(\mathbf{r}) + \frac{1}{8\pi a^2} \frac{e^{-(r/a)}}{r} \right\}, \qquad \langle B^2 \rangle_{\mathbf{r}} = 8\pi T \delta(\mathbf{r}). \qquad (4.21)$$

REFERENCE

1. A. I. Akhiezer and A. G. Sitenko, "On the Theory of Electromagnetic Fluctuations in Plasmas." FTI AN USSR, Khar'kov, 1960.

5
Taking Account of Ion Motion. Fluctuations in an Electron-Ion Plasma

---

1. Dielectric Permittivities of a Plasma Taking Account of Ion Motion

Up to now we have considered the electromagnetic properties of a plasma, as well as the fluctuations in the plasma, without taking account of the ion motion. In particular, formulas (3.21) and (3.22) for the dielectric permittivities ϵ_l and ϵ_t of the plasma have been obtained without taking account of the ion motion. In reality, however, the ion motion affects the electromagnetic properties of a plasma, particularly in the low-frequency range, and leads to an essential change in the nature of the fluctuations in the plasma.

Because of the large difference between the electron and ion masses, the energy exchange between the electrons and ions occurs very slowly, hence, a quasi-equilibrium state is possible in the plasma, for which the electrons and ions are characterized by different temperatures T_e and T_i. Such a nonisothermal plasma possesses a number of specific peculiarities.

Let us assume that an equilibrium or quasi-equilibrium plasma is electrically neutral. Let δn^e and δn^i denote the deviations of the electron and ion densities from the equilibrium value n_0. The charge and current densities in the plasma will hence be determined be the expressions:

$$\rho = e(\delta n^e - \delta n^i), \qquad \mathbf{j} = \mathbf{j}^e + \mathbf{j}^i, \tag{5.1}$$

$\mathbf{j}^e$ and $\mathbf{j}^i$ are the electron and ion components of the current connected with the densities δn^e and δn^i by means of the continuity relationships

$$e \frac{\partial \delta n^e}{\partial t} + \operatorname{div} \mathbf{j}^e = 0, \qquad - e \frac{\partial \delta n^i}{\partial t} + \operatorname{div} \mathbf{j}^i = 0. \tag{5.2}$$

Just as the total current $\mathbf{j}$, the electron and ion components of the current $\mathbf{j}^e$ and $\mathbf{j}^i$ are linearly related to the electric field intensity in the plasma $\mathbf{E}$

$$j_i^\alpha = \chi_{ij}^\alpha \dot{E}_j , \tag{5.3}$$

where the superscript α denotes the kind of particles (e or i), χ_{ij}^e and χ_{ij}^i are the electron and ion electrical susceptibilities of the plasma. The dielectric permittivity of the plasma ϵ_{ij} is associated with the electron and ion susceptibilities χ_{ij}^e and χ_{ij}^i by means of the relationships

$$\epsilon_{ij} = \delta_{ij} + 4\pi \sum_\alpha \chi_{ij}^\alpha . \tag{5.4}$$

In order to find the electrical susceptibilities χ_{ij}^e and χ_{ij}^i , let us note that in neglecting interaction between particles the electron and ion fluctuations in the plasma are independent. Hence, on the basis of the inversion of the fluctuation–dissipation theorem (2.71), it is possible to write

$$\chi_{ij}^\alpha = -\frac{1}{4\pi} \left\{ \frac{\Omega_\alpha^2}{\omega^2} \delta_{ij} + \frac{2}{\omega^2} \int_{-\infty}^{\infty} \frac{\omega^0 (\partial/\partial E) \langle j_i^\alpha j_i^\alpha \rangle_{\mathbf{k}\omega'}^0}{\omega' - \omega - io} \, d\omega' \right\} \tag{5.5}$$

for each component of the electrical susceptibility χ_{ij}^α , where $\langle j_i^\alpha j_j^\alpha \rangle_{\mathbf{k}\omega}^0$ is the spectral distribution of the current fluctuations of the free particles (ions or electrons). Assuming no external fields, formula (3.17) may be used for the spectral distributions of the electron and ion current fluctuations:

$$\langle j_i^\alpha j_j^\alpha \rangle_{\mathbf{k}\omega}^0 = 2\pi e^2 n_0 \int d\mathbf{v} \, v_i v_j f_0^\alpha(v) \, \delta(\omega - \mathbf{k}\mathbf{v}) \tag{5.6}$$

where $f_0^\alpha(v)$ is the electron or ion distribution function in the plasma. We hence obtain the dielectric permittivity tensor of the plasma as

$$\epsilon_{ij}(\omega, \mathbf{k}) = \delta_{ij} - \sum_\alpha \frac{\Omega_\alpha^2}{\omega^2} \left\{ \delta_{ij} + \int \frac{v_i v_j \mathbf{k} \dfrac{\partial f_0^\alpha(v)}{\partial \mathbf{v}}}{\mathbf{kv} - \omega - io} \, d\mathbf{v} \right\}. \tag{5.7}$$

According to (5.7), the ion contribution to ϵ_{ij} is expressed by terms of the same kind as for the electrons, but just with the corresponding change in mass and distribution function.

In the isotropic case it is convenient to separate the longitudinal and transverse parts ϵ_l and ϵ_t out of ϵ_{ij}. Selecting the Maxwell functions with temperatures T_e and T_i as the electron and ion distribution functions, we obtain the following expressions for the dielectric permittivity of the plasma:

$$\epsilon_l(\omega, \mathbf{k}) = 1 + \frac{1}{a^2 k^2} \{1 - \varphi(z) + t[1 - \varphi(\mu z)]$$
$$+ i \sqrt{\pi} z (e^{-z^2} + t\mu e^{-\mu^2 z^2})\}, \tag{5.8}$$

$$\epsilon_t(\omega, \mathbf{k}) = 1 - \frac{\Omega^2}{\omega^2} \left\{ \varphi(z) + \frac{t}{\mu} \varphi(\mu z) - i \sqrt{\pi} z \left(e^{-z^2} + \frac{t}{\mu} e^{-\mu^2 z^2} \right) \right\}, \tag{5.9}$$

where the function $\varphi(z)$ is determined by (3.26), $z = \sqrt{\frac{3}{2}} (\omega/ks)$, $t = T_e/T_i$, and $\mu^2 = (M/m)(T_e/T_i)$ (M is the mass of the ion).

Let us also present the formulas for the electron and ion components of the longitudinal and transverse electrical susceptibilities of the plasma:

$$\chi_l^e(\omega, \mathbf{k}) = \frac{1}{4\pi a^2 k^2} \{1 - \varphi(z) + i \sqrt{\pi} z \, e^{-z^2}\},$$

$$\chi_l^i(\omega, \mathbf{k}) = \frac{t}{4\pi a^2 k^2} \{1 - \varphi(\mu z) + i \sqrt{\pi} \mu z \, e^{-\mu^2 z^2}\}, \tag{5.10}$$

$$\chi_t^e(\omega, \mathbf{k}) = -\frac{1}{4\pi} \frac{\Omega^2}{\omega^2} \{\varphi(z) - i \sqrt{\pi} z \, e^{-z^2}\},$$

$$\chi_t^i(\omega, \mathbf{k}) = -\frac{t}{4\pi\mu} \frac{\Omega^2}{\omega^2} \{\varphi(\mu z) - i \sqrt{\pi} z \, e^{-\mu^2 z^2}\}. \tag{5.11}$$

Let us note that because of the smallness of the ratio between the electron and ion masses $t/\mu^2 \equiv m/M$ the ion component of the

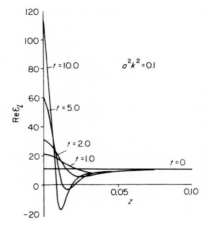

FIG. 10.

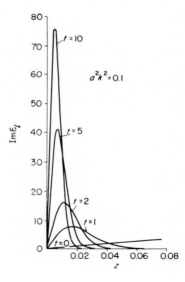

FIG. 11.

transverse electric susceptibility of the plasma is very small. Hence, in practice, the ion motion may always be neglected in the transverse dielectric permittivity of the plasma. The real and imaginary parts of the longitudinal dielectric permittivity of the plasma as a function of the nondimensional frequency z are presented in Figs. 10 and 11 for different values of the temperature $t = T_e/T_i$.

2. Longitudinal Waves in an Electron–Ion Plasma

The expressions obtained for ϵ_l and ϵ_t taking account of ion motion completely determine the electromagnetic properties of the plasma. In particular, knowing ϵ_l, it is easy to clarify what influence the ion motion exerts on the longitudinal oscillations in the plasma. Using (5.8), the dispersion equation for the longitudinal waves in an electron–ion plasma may be written as

$$1 + \frac{1}{a^2k^2}\{1 - \varphi(z) + t[1 - \varphi(\mu z)] + i\sqrt{\pi}z(e^{-z^2} + t\mu\, e^{-\mu^2z^2})\} = 0.$$

$$(5.12)$$

Weakly-damped oscillations in a plasma, which we will consider, are of greatest interest.

As is easy to see, the influence of ion motion on longitudinal oscillations may be neglected in the high-frequency range. In fact, if the phase velocity of the waves ω/k significantly exceeds the thermal electron velocity s ($z \gg 1$), then the contribution of the ion terms in (5.12) is negligible, since under ordinary conditions $\mu \gg 1$. Hence, the frequency and damping coefficient of the high-frequency plasma waves associated with the oscillations of the charge density in the plasma, are determined, in practice, just by the electron motion. In the low-frequency range, for which the phase velocities of the waves are less than the thermal electron velocities, the role of ions becomes essential, particularly in the case of a nonisothermal plasma. Assuming that the phase velocity of the waves ω/k is considerably less than the thermal electron velocity s, but considerably greater than the mean thermal velocity of the ions s^i, the dispersion equation (5.12) may be represented approximately as

$$1 + \frac{1}{a^2k^2}\left\{1 - \frac{t}{2\mu^2z^2} + i\sqrt{\pi}z\right\} = 0, \qquad \frac{1}{\mu} \ll z \ll 1. \qquad (5.13)$$

We have used the expansion (3.27) for $\varphi(z)$ and the expansion (3.28) for $\varphi(\mu z)$. Considering k real, in a first approximation we obtain the following expressions for the frequency and damping coefficient from (5.13) (1, 2):

$$\omega = \frac{kv_s}{(1 + a^2k^2)^{1/2}},$$
(5.14)

$$\gamma = \left(\frac{\pi}{8}\frac{m}{M}\right)^{1/2}\frac{\omega}{(1 + a^2k^2)^{3/2}},$$
(5.15)

where $v_s = (T_e/M)^{1/2}$.

If $ak \ll 1$, the frequency (5.14) turns out to be proportional to the wave vector $\mathbf{k}$, and the corresponding oscillations are in the nature of sonic waves

$$\omega = kv_s.$$
(5.16)

Here the speed of sound in the plasma v_s is determined by the electron temperature T_e and the mass M of the ions. Substituting the frequency (5.16) into the condition $\mu z \gg 1$ it is easy to see that this condition will be satisfied only if $T_e \gg T_i$, i.e., weakly damped sonic oscillations are possible only in a strongly non-isothermal plasma.

In the opposite limiting case $ak \gg 1$, the frequency (5.14) turns out to be independent of the wave vector and characterizes the so-called ion oscillations of the plasma

$$\omega = \Omega_i \equiv \left(\frac{4\pi e^2 n_0}{M}\right)^{1/2}.$$
(5.17)

Damping of the ion oscillations is determined by (5.15) only when $a^i k \ll 1$ (for $a^i k \gg 1$ the damping increases sharply). It follows from the inequalities $a^i k \ll 1 \ll ak$ that, exactly as weakly damped sonic oscillations, weakly damped ion oscillations are possible only in a strongly nonisothermal plasma $T_e \gg T_i$.

Let us note that if the high-frequency Langmuir oscillations in a plasma are accompanied by charge-density oscillations, then the low-frequency ion and sonic oscillations are accompanied only by mass-density oscillations, since the nonequilibrium electron-charge density is cancelled in the low-frequency oscillations by the nonequilibrium ion-charge density.

3. Fluctuations in Charge and Current Densities in an Equilibrium Electron–Ion Plasma

If an electron–ion plasma is in a state of complete statistical equilibrium (the electron and ion temperatures are equal $T_e = T_i = T$), the charge and current density fluctuations, as well as the fields, in the plasma will be determined by the general formulas obtained in Chapter 4 [for example, by formulas (4.1), (4.10), and (4.16)]. The expressions (5.8) and (5.9), in which we should put $t = 1$, should be used here for the dielectric permittivities ϵ_l and ϵ_t.

Since the corrections to the transverse dielectric permittivity which are associated with the ion motion are negligibly small, taking account of the ion motion does not, in practice, affect the fluctuations of the transverse current density and of the transverse

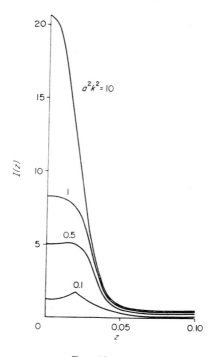

Fig. 12.

fields in the plasma. However, the ion motion affects the fluctuations of the longitudinal current density and of the charge density in the plasma. Substituting (5.8) into (4.1), we obtain the following general formula for the spectral distribution of the charge-density fluctuations in the plasma, taking ion motion into account:

$$\langle \rho^2 \rangle_{k\omega} = \sqrt{2\pi}\, e^2 n_0 \frac{a^3 k^3}{\Omega} \frac{e^{-z^2} + \mu e^{-\mu^2 z^2}}{[a^2 k^2 + 2 - \varphi(z) - \varphi(\mu z)]^2 + \pi z^2 (e^{-z^2} + \mu e^{-\mu^2 z^2})^2}, \tag{5.18}$$

where $z = \sqrt{\tfrac{3}{2}}\,(\omega/ks)$, $\mu^2 = M/m$ and $a^2 = T/4\pi e^2 n_0$.

Integrating $\langle \rho^2 \rangle_{k\omega}$ with respect to the frequencies, which is done most simply by using the Kramers–Kronig relationships (2.47), we find the Fourier components of the spatial correlation function

$$\langle \rho^2 \rangle_k = \frac{e^2 n_0 k^2}{(k^2/2) + (4\pi e^2 n_0/T)}. \tag{5.19}$$

This formula may be obtained from (4.2) by replacing n_0 by $2n_0$. Let us note that the ion contribution to $\langle \rho^2 \rangle_k$ is particularly essential if $a^2 k^2 > 1$ (for $a^2 k^2 \gg 1$ the magnitude of $\langle \rho^2 \rangle_k$ is doubled as compared with the electron plasma case), the ion contribution to $\langle \rho^2 \rangle_k$ decreases as $a^2 k^2$ decreases. The correlation function for the charge density fluctuations is determined by the previous formula (4.3) in which n_0 should be replaced by $2n_0$:

$$\langle \rho^2 \rangle_r = 2e^2 n_0 \left\{ \delta(\mathbf{r}) - \frac{1}{4\pi a^2} \frac{\exp[-\sqrt{2}(r/a)]}{r} \right\}. \tag{5.20}$$

The dependence of $I(z) = 2\langle \rho^2 \rangle_{kz}/\langle \rho^2 \rangle_k$ on z is shown in Fig. 12 for different values of $a^2 k^2$ for a plasma containing hydrogen ions. We see that for small z the curve of $I(z)$, compared with the curve in Fig. 9, has an additional maximum corresponding to the presence of low-frequency ion oscillations. For large z, the influence of the ions on the spectral distribution turns out to be insignificant.

4. Electron and Ion Density Fluctuations in an Equilibrium Plasma

In addition to charge-density fluctuations, independent fluctuations of the electron and ion densities are also possible in an electron–ion plasma. To find these fluctuations it is necessary to introduce random fields acting independently on the electrons and ions, into the material equations (5.3). Then, using the Maxwell equations and the relationship (5.1), it is easy to find the proportionality coefficients between the densities of the electron and ion currents and the corresponding random potentials. These coefficients will be expressed not only in terms of the plasma dielectric permittivities ϵ_l and ϵ_t, but also in terms of the electron and ion susceptibilities of the plasma χ_l^e, χ_l^i, χ_t^e, and χ_t^i. By using the fluctuation–dissipation theorem, we obtain the following formulas for the spectral distributions of the electron and ion current densities:

$$\langle j_i^e j_j^e \rangle_{\mathbf{k}\omega} = 2T\omega \left\{ \frac{k_i k_j}{k^2} \operatorname{Im} \frac{\chi_l^e(1 + 4\pi\chi_l^i)}{\epsilon_l} \right.$$
$$\left. + \left(\delta_{ij} - \frac{k_i k_j}{k^2} \right) \operatorname{Im} \frac{\chi_t^e(1 + 4\pi\chi_t^i - \eta^2)}{\epsilon_t - \eta^2} \right\}, \qquad (5.21)$$

$$\langle j_i^i j_j^i \rangle_{\mathbf{k}\omega} = 2T\omega \left\{ \frac{k_i k_j}{k^2} \operatorname{Im} \frac{\chi_l^i(1 + 4\pi\chi_l^e)}{\epsilon_l} \right.$$
$$\left. + \left(\delta_{ij} - \frac{k_i k_j}{k^2} \right) \operatorname{Im} \frac{\chi_t^i(1 + 4\pi\chi_t^e - \eta^2)}{\epsilon_t - \eta^2} \right\}. \qquad (5.22)$$

Making use of the continuity equation for the electrons and ions, which connects the Fourier components of the particle densities and the currents, the following formulas for the spectral distributions of the electron and ion density fluctuations are easily obtained from (5.21) and (5.22):

$$\langle (\delta n^e)^2 \rangle_{\mathbf{k}\omega} = 8\pi n_0 \frac{a^2 k^2}{\omega} \operatorname{Im} \frac{\chi_l^e(1 + 4\pi\chi_l^i)}{\epsilon_l}, \qquad (5.23)$$

$$\langle (\delta n^i)^2 \rangle_{\mathbf{k}\omega} = 8\pi n_0 \frac{a^2 k^2}{\omega} \operatorname{Im} \frac{\chi_l^i(1 + 4\pi\chi_l^e)}{\epsilon_l}. \qquad (5.24)$$

Substituting (5.10) and (5.8) for χ_l^e, χ_l^i, and ϵ_l, the spectral distributions of the electron and ion density fluctuations are respresentable as

$$\langle (\delta n^e)^2 \rangle_{\mathbf{k}\omega} = \sqrt{6\pi} \frac{n_0}{ks} \frac{\{[a^2 k^2 + 1 - \varphi(\mu z)]^2 + \pi\mu^2 z^2 e^{-2\mu^2 z^2}\}e^{-z^2}}{[a^2 k^2 + 2 - \varphi(z) - \varphi(\mu z)]^2 + \pi z^2 (e^{-z^2} + \mu e^{-\mu^2 z^2})^2} \begin{array}{c} \\ + \mu\{[1 - \varphi(z)]^2 + \pi z^2 e^{-2z^2}\}e^{-\mu^2 z^2} \end{array}$$

$$(5.25)$$

$$\langle (\delta n^i)^2 \rangle_{\mathbf{k}\omega} = \sqrt{6\pi}\, \frac{n_0}{ks} \frac{\begin{aligned}\mu\{[a^2k^2 + 1 - \varphi(z)]^2 + \pi z^2 e^{-2z^2}\}e^{-\mu^2 z^2} \\ + \{[1 - \varphi(\mu z)]^2 + \pi \mu^2 z^2 e^{-2\mu^2 z^2}\}e^{-z^2}\end{aligned}}{[a^2k^2 + 2 - \varphi(z) - \varphi(\mu z)]^2 + \pi z^2(e^{-z^2} + \mu e^{-\mu^2 z^2})^2}.$$

$$(5.26)$$

Integrating (5.23) and (5.24) with respect to the frequencies by using (2.47), we have

$$\langle (\delta n^e)^2 \rangle_{\mathbf{k}} = n_0 \left. \frac{1 + a^2k^2 - \varphi(\mu z)}{2 + a^2k^2 - \varphi(\mu z)} \right|_{z \to 0},$$

$$\langle (\delta n^i)^2 \rangle_{\mathbf{k}} = n_0 \left. \frac{(1 + a^2k^2)\,[1 - \varphi(\mu z)]}{2 + a^2k^2 - \varphi(\mu z)} \right|_{z \to 0}.$$

$$(5.27)$$

If the ions are considered to be infinitely heavy ($\mu \to \infty$), then we find by virtue of $\varphi(\mu z)|_{\mu \to \infty} = 1$:

$$\langle (\delta n^e)^2 \rangle_{\mathbf{k}} = n_0 \frac{a^2k^2}{1 + a^2k^2}, \qquad \langle (\delta n^i)^2 \rangle_{\mathbf{k}} = 0. \qquad (5.28)$$

In this case the charge density fluctuations in the plasma are determined by the electron density fluctuations (the ions are fixed).

In reality, the ions have a finite mass, hence $\varphi(\mu z)|_{z \to 0} = 0$, and therefore (3, 4)

$$\langle (\delta n^e)^2 \rangle_{\mathbf{k}} = \langle (\delta n^i)^2 \rangle_{\mathbf{k}} = n_0 \frac{1 + a^2k^2}{2 + a^2k^2}. \qquad (5.29)$$

If $a^2k^2 \gg 1$, the Coulomb interaction between particles may be neglected and (5.29) leads to the well-known result for the mean-square fluctuations of the density of neutral particles:

$$\langle (\delta n^e)^2 \rangle_{\mathbf{k}} = \langle (\delta n^i)^2 \rangle_{\mathbf{k}} = n_0. \qquad (5.30)$$

In the limiting case $a^2k^2 \ll 1$ the mean-square fluctuations turn out to be half of (5.30):

$$\langle (\delta n^e)^2 \rangle_{\mathbf{k}} = \langle (\delta n^i)^2 \rangle_{\mathbf{k}} = \tfrac{1}{2} n_0. \qquad (5.31)$$

5. Fluctuations in a Nonisothermal Electron–Ion Plasma (Isotropic Case)

Let us now consider the fluctuations in an electron–ion plasma in which the electrons and ions are characterized by different temperatures T_e and T_i.* Due to the great difference in the masses, the energy exchange between the electrons and ions proceeds considerably more slowly than between particles of the same sort, hence, a nonisothermal plasma may be considered as a quasi-equilibrium system, and general methods may be used to investigate fluctuations therein. It must, however, be kept in mind that the electrons and ions in the plasma are bound together by the self-consistent field, hence, the charge and current density fluctuations of the electron and ion components in the plasma are not independent, and in particular, will be determined by both the plasma temperatures T_e and T_i.

Let us consider in detail the derivation of the fundamental relationships for fluctuations in a nonisothermal plasma. Let us introduce the secondary random fields $\tilde{\mathbf{E}}_i$ and $\tilde{\mathbf{E}}_e$, acting independently on the electrons and ions, into the material equation (5.3):

$$j_i^\alpha = - i\omega\chi_{ij}^\alpha (E_j + \tilde{E}_j^\alpha), \qquad \alpha = e, i. \tag{5.32}$$

The self-consistent field $\mathbf{E}$ is determined by the Maxwell equations

$$\Lambda_{ij}^0 E_j = \frac{4\pi}{i\omega} (j_i^e + j_i^i). \tag{5.33}$$

Eliminating the self-consistent electric field $\mathbf{E}$ from (5.32) by using the Maxwell equations (5.33), in the case of an isotropic plasma we easily obtain the following relationships between the electron and ion current densities $\mathbf{j}^\alpha$ and the random potentials $\tilde{\mathbf{A}}^\alpha$ ($\tilde{\mathbf{E}}^\alpha = i\omega\tilde{\mathbf{A}}^\alpha$):

$$\mathbf{j}_l^e = \omega^2 \frac{\chi_l^e}{\epsilon_l^e} \{(1 + 4\pi\chi_l^i) \tilde{\mathbf{A}}_l^e - 4\pi\chi_l^i\tilde{\mathbf{A}}_l^i\},$$
$$\mathbf{j}_l^i = \omega^2 \frac{\chi_l^i}{\epsilon_l} \{-4\pi\chi_l^e\tilde{\mathbf{A}}_l^e + (1 + 4\pi\chi_l^e) \tilde{\mathbf{A}}_l^i\}, \tag{5.34}$$

---

* Akhiezer, Akhiezer, and Sitenko (4) considered fluctuations in a nonisothermal plasma on the basis of the kinetic equations.

$$\mathbf{j}_t^e = \omega^2 \frac{\chi_t^e}{\epsilon_t - \eta^2} \{(1 + 4\pi\chi_t^i - \eta^2) \tilde{\mathbf{A}}_t^e - 4\pi\chi_t^i \tilde{\mathbf{A}}_t^i\},$$

$$\mathbf{j}_t^i = \omega^2 \frac{\chi_t^i}{\epsilon_t - \eta^2} \{- 4\pi\chi_t^e \tilde{\mathbf{A}}_t^e + (1 + 4\pi\chi_t^e - \eta^2) \tilde{\mathbf{A}}_t^i\}.$$

(5.35)

The presence of the self-consistent field leads to the fact that the electron current turns out to be dependent on the ion secondary potential, and the ion current, in turn, depends on the electron secondary potential.

Since there is a connection between the currents $\mathbf{j}^e$ and $\mathbf{j}^i$, and the system is not in a state of total equilibrium, it is then impossible to determine the fluctuations directly by means of the coefficients of the proportionality between $\mathbf{j}^\alpha$ and $\tilde{\mathbf{A}}^\alpha$, according to (1.25). However, by using the relationships (5.34) and (5.35), the current correlators may be expressed in terms of the secondary potential $\tilde{\mathbf{A}}^\alpha$ correlators. According to the assumption, the secondary potentials $\tilde{\mathbf{A}}^e$ and $\tilde{\mathbf{A}}^i$ are independent, hence, the fluctuations of each of them may be determined only by the temperature of the appropriate subsystem (the electron or ion temperature). Solving the system (5.34) and (5.35) for the potentials we find

$$\omega^2 \tilde{\mathbf{A}}_l^\alpha = \frac{1}{\chi_l^\alpha} \mathbf{j}_l^\alpha + 4\pi\mathbf{j}_l, \qquad \omega^2 \tilde{\mathbf{A}}_t^\alpha = \frac{1}{\chi_t^\alpha} \mathbf{j}_t^\alpha + \frac{4\pi}{1 - \eta^2} \mathbf{j}_t, \qquad \alpha = e, i.$$

(5.36)

Then using the fluctuation–dissipation theorem in the form (1.27), we obtain the following formulas for the spectral distribution of the fluctuations of the secondary potentials:

$$\langle A_i^\alpha A_j^\alpha \rangle_{\mathbf{k}\omega} = \frac{2T^\alpha}{\omega^3} \left\{ \frac{k_i k_j}{k^2} \frac{\operatorname{Im}\chi_l^\alpha}{|\chi_l^\alpha|^2} + \left(\delta_{ij} - \frac{k_i k_j}{k^2}\right) \frac{\operatorname{Im}\chi_t^\alpha}{|\chi_t^\alpha|^2} \right\}, \qquad \alpha = e, i,$$

(5.37)

$$\langle A_i^e A_j^i \rangle_{\mathbf{k}\omega} = 0.$$

(5.38)

As should have been expected, the fluctuations in the electron potential are expressed in terms of the electron susceptibility of the plasma, and the fluctuations in the ion potential, in terms of the ion susceptibility of the plasma. By using the relationships (5.34) and (5.35), we obtain the following general formulas for the spectral distributions of the current fluctuations in a noniso-thermal plasma:

$$\langle j_i^e j_j^e \rangle_{\mathbf{k}\omega} = \frac{2\omega}{|\epsilon_l|^2} \frac{k_i k_j}{k^2} \{ T_e \, | \, 1 + 4\pi\chi_l^i \, |^2 \, \mathrm{Im} \, \chi_l^e + T_i \, | \, 4\pi\chi_l^e \, |^2 \, \mathrm{Im} \, \chi_l^i \}$$

$$+ \frac{2\omega}{|\epsilon_t - \eta^2|^2} \left(\delta_{ij} - \frac{k_i k_j}{k^2} \right)$$

$$\times \{ T_e \, | \, 1 + 4\pi\chi_t^i - \eta^2 \, |^2 \, \mathrm{Im} \, \chi_t^e + T_i \, | \, 4\pi\chi_t^e \, |^2 \, \mathrm{Im} \, \chi_t^i \},$$

$$(5.39)$$

$$\langle j_i^i j_j^i \rangle_{\mathbf{k}\omega} = \frac{2\omega}{|\epsilon_l|^2} \frac{k_i k_j}{k^2} \{ T_e \, | \, 4\pi\chi_l^i \, |^2 \, \mathrm{Im} \, \chi_l^e + T_i \, | \, 1 + 4\pi\chi_l^e \, |^2 \, \mathrm{Im} \, \chi_l^i \}$$

$$+ \frac{2\omega}{|\epsilon_t - \eta^2|^2} \left(\delta_{ij} - \frac{k_i k_j}{k^2} \right)$$

$$\times \{ T_e \, | \, 4\pi\chi_t^i \, |^2 \, \mathrm{Im} \, \chi_t^e + T_i \, | \, 1 + 4\pi\chi_t^e - \eta^2 |^2 \, \mathrm{Im} \, \chi_t^i \},$$

$$(5.40)$$

$$\langle j_i^e j_j^i \rangle_{\mathbf{k}\omega} = \langle j_i^i j_j^e \rangle_{\mathbf{k}\omega}^* = - \frac{2\omega}{|\epsilon_l|^2} \frac{k_i k_j}{k^2} \{ T_e(1 + 4\pi\chi_l^i) \, 4\pi\chi_l^{i*} \, \mathrm{Im} \, \chi_l^e$$

$$+ T_i 4\pi\chi_l^e(1 + 4\pi\chi_l^{e*}) \, \mathrm{Im} \, \chi_l^i \}$$

$$- \frac{2\omega}{|\epsilon_t - \eta^2|^2} \left(\delta_{ij} - \frac{k_i k_j}{k^2} \right) \{ T_e(1 + 4\pi\chi_t^i - \eta^2) \, 4\pi\chi_t^{i*} \, \mathrm{Im} \, \chi_t^e$$

$$+ T_i 4\pi\chi_t^e(1 + 4\pi\chi_t^{e*} - \eta^2) \, \mathrm{Im} \, \chi_t^i \}.$$

$$(5.41)$$

By using (5.39)–(5.41) we easily find the spectral distributions of the fluctuations of any other quantities in a nonisothermal plasma. Thus, we obtain the following expressions for the spectral distributions of the electron and ion density fluctuations in a two-temperature plasma:

$$\langle (\delta n^e)^2 \rangle_{\mathbf{k}\omega} = \frac{8\pi n_0 a^2 k^2}{\omega \, |\epsilon_l|^2} \left\{ | \, 1 + 4\pi\chi_l^i \, |^2 \, \mathrm{Im} \, \chi_l^e + \frac{T_i}{T_e} \, | \, 4\pi\chi_l^e \, |^2 \, \mathrm{Im} \, \chi_l^i \right\},$$

$$(5.42)$$

$$\langle (\delta n^i)^2 \rangle_{\mathbf{k}\omega} = \frac{8\pi n_0 a^2 k^2}{\omega \, |\epsilon_l|^2} \left\{ | \, 4\pi\chi_l^i \, |^2 \, \mathrm{Im} \, \chi_l^e + \frac{T_i}{T} \, | \, 1 + 4\pi\chi_l^e \, |^2 \, \mathrm{Im} \, \chi_l^i \right\},$$

$$(5.43)$$

$$\langle \delta n^e \delta n^i \rangle_{\mathbf{k}\omega} = \langle \delta n^i \delta n^e \rangle_{\mathbf{k}\omega}^* = \frac{8\pi n_0 a^2 k^2}{\omega \mid \epsilon_l \mid^2} \left\{ (1 + 4\pi\chi_l^{~i}) \, 4\pi\chi_l^{i*} \, \mathrm{Im} \, \chi_l^{~e} \right.$$

$$\left. + \frac{T_i}{T_e} 4\pi\chi_l^{~e}(1 + 4\pi\chi_l^{e*}) \, \mathrm{Im} \, \chi_l^{~i} \right\}. \tag{5.44}$$

The dependence of the spectral distributions $2\langle(\delta n^e)^2\rangle_{\mathbf{k}z}/n_0$ and $2\langle(\delta n^i)^2\rangle_{\mathbf{k}z}/n_0$ on the nondimensional frequency z is presented in Figs. 13 and 14 for various values of the parameters $a^2 k^2 = 0.1$ and $t = 1; 2; 5;$ and 10.

Utilizing (5.42)–(5.44), we may find the spectral distribution of the charge-density fluctuations in a two-temperature plasma:

$$\langle \rho^2 \rangle_{\mathbf{k}\omega} = e^2 \langle (\delta n^e - \delta n^i)^2 \rangle = \frac{8\pi e^2 n_0 a^2 k^2}{\omega \mid \epsilon_l \mid^2} \, \mathrm{Im} \left\{ \chi_l^{~e} + \frac{T_i}{T_e} \chi_l^{~i} \right\}. \tag{5.45}$$

This distribution may be represented as

$$\langle \rho^2 \rangle_{\mathbf{k}\omega} = \sqrt{2\pi} \, e^2 n_0 \, \frac{a^3 k^3}{\Omega}$$

$$\times \frac{e^{-z^2} + \mu e^{-\mu^2 z^2}}{\{a^2 k^2 + 1 - \varphi(z) + t[1 - \varphi(\mu z)]\}^2 + \pi z^2 (e^{-z^2} + t\mu e^{-\mu^2 z^2})^2}. \tag{5.46}$$

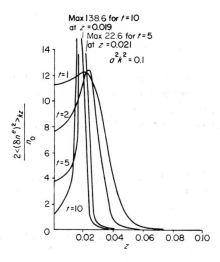

Fɪɢ. 13.

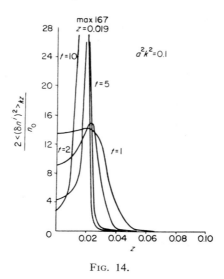

FIG. 14.

Presented in Fig. 15 is the dependence of $2\langle\rho^2\rangle_{kz}/e^2 n_0$ on the nondimensional frequency z for $a^2 k^2 = 0.1$ and $t = 1; 2; 5;$ and 10. Let us present approximate formulas for the spectral distribution of the charge-density fluctuations in various limiting cases for $a^2 k^2 \ll 1$. In the low-frequency domain $\omega \ll k s^i$ $(z \ll \mu^{-1})$ we have

$$\langle\rho^2\rangle_{k\omega} \cong \sqrt{2\pi}\, e^2 n_0 \frac{a^3 k^3}{\Omega} \frac{\mu}{(1 + t)^2}. \tag{5.47}$$

As the frequency increases, $\langle\rho^2\rangle_{k\omega}$ will decrease exponentially. If $z \lesssim \mu^{-1} \ll 1$, then

$$\langle\rho^2\rangle_{k\omega} \simeq \sqrt{2\pi}\, e^2 n_0 \frac{a^3 k^3}{\Omega} \frac{\mu e^{-\mu^2 z^2}}{\{1 + t[1 - \varphi(\mu z)]\}^2 + \pi\mu^2 z^2 e^{-2\mu^2 z^2}}. \tag{5.48}$$

In order of magnitude the width of the maximum in the low-frequency domain equals $z \sim \mu^{-1}$. According to (5.47), the height of the maximum depends strongly on the degree of isothermy of the plasma. In the case of a strongly nonisothermal plasma $(t \gg 1)$ the height of the maximum (5.47) is significantly reduced as compared with the isothermal case.

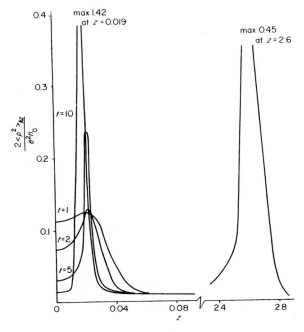

FIG. 15.

If $\mu^{-1} \ll z \ll 1$, then

$$\langle \rho^2 \rangle_{\mathbf{k}\omega} \simeq \sqrt{2\pi}\, e^2 n_0 \frac{a^3 k^3}{\Omega} \frac{1 + \mu e^{-\mu^2 z^2}}{\left(a^2 k^2 + 1 - \dfrac{t}{2\mu^2 z^2}\right)^2 + \pi z^2 (1 + t\mu e^{-\mu^2 z^2})^2}.$$
(5.49)

In this frequency range the spectral density $\langle \rho^2 \rangle_{\mathbf{k}\omega}$ for the isothermal plasma $(t = 1)$ is very small [μ times less than the value at the maximum (5.47)]. In a strongly nonisothermal plasma $(t \gg 1)$ the spectral density of the fluctuations has sharp maximums at $\omega = \pm\omega_s(k)$, where $\omega_s(k) = kv_s$ is the frequency of the non-isothermal sonic oscillations:

$$\langle \rho^2 \rangle_{\mathbf{k}\omega} \simeq \pi e^2 n_0 a^4 k^4 \{\delta(\omega - kv_s) + \delta(\omega + kv_s)\}.$$
(5.50)

Later, as the frequency increases $\omega \sim ks\,(z \sim 1)$ the spectral density of the fluctuations decreases exponentially $\sim e^{-z^2}$. At high frequencies $\omega \gg ks \, (z \gg 1)$ electrons play a basic role in the charge-density fluctuations, and $\langle \rho^2 \rangle_{\mathbf{k}\omega}$ is determined by a formula agreeing with (4.5). In particular, $\langle \rho^2 \rangle_{\mathbf{k}\omega}$ has maximums at frequencies corresponding to the plasma oscillations. In the limiting case of short wavelengths $a^2 k^2 \gg 1$ formula (5.46) takes the form:

$$\langle \rho^2 \rangle_{\mathbf{k}\omega} \simeq \sqrt{2\pi} \, \frac{e^2 n_0}{\Omega a k} \, (e^{-z^2} + \mu e^{-\mu^2 z^2}). \tag{5.51}$$

Let us present the formula for the spectral distribution of the electric field fluctuations in a nonisothermal plasma:

$$\langle E_i E_j \rangle_{\mathbf{k}\omega} = \frac{32\pi^2}{\omega} \left\{ \frac{k_i k_j}{k^2} \, \frac{1}{|\epsilon_l|^2} \, (T_e \operatorname{Im} \chi_l^{\,e} + T_i \operatorname{Im} \chi_l^{\,i}) \right.$$

$$\left. + \left(\delta_{ij} - \frac{k_i k_j}{k^2} \right) \frac{1}{|\epsilon_t - \eta^2|^2} (T_e \operatorname{Im} \chi_t^{\,e} + T_i \operatorname{Im} \chi_t^{\,i}) \right\}. \tag{5.52}$$

6. Fluctuations in a Nonisothermal Electron–Ion Plasma (Anisotropic Case)

Now, let us generalize the obtained results in the case of an anisotropic nonisothermal plasma. In particular, this generalization will be applicable for a nonisothermal plasma in an external permanent and homogeneous magnetic field $\mathbf{B_0}$.

Introducing the secondary field in the material equations (5.32) and utilizing the Maxwell equations (5.33), we obtain the self-consistent electric field in the form

$$E_j = -4\pi i \omega \Lambda_{ji}^{-1} \{ \chi_{ik}^e \tilde{A}_k^{\,e} + \chi_{ik}^i \tilde{A}_k^{\,i} \}, \tag{5.53}$$

where

$$\Lambda_{ij} = \eta^2 \left(\frac{k_i k_j}{k^2} - \delta_{ij} \right) + \epsilon_{ij}$$

(just as χ_{ij}^e and χ_{ij}^i in the anisotropic case, the dielectric permittivity tensor of the plasma ϵ_{ij} does not reduce to longitudinal and transverse components). Substituting (5.53) into (5.32) we find the relation expressing the current in terms of the secondary potentials:

$$j_i^{\alpha} = \omega^2 \sum_{\beta} (\delta_{ij}\delta_{\alpha\beta} - 4\pi\chi_{ik}^{\alpha}\Lambda_{kj}^{-1}) \chi_{jl}^{\beta}\tilde{A}_l^{\beta}. \tag{5.54}$$

Conversely, the potentials $\tilde{A}_i^{\alpha}$ are expressed in terms of the currents j_j^{α} by means of the equalities

$$\omega^2 \tilde{A}_i^{\alpha} = (\chi_{ij}^{\alpha})^{-1} j_j^{\alpha} + 4\pi(\Lambda_{ij}^0)^{-1} j_j. \tag{5.55}$$

Remarking that Λ_{ij}^0 is a real symmetric tensor, we find on the basis of the fluctuation–dissipation formula (1.27)

$$\langle A_i^{\alpha} A_j^{\alpha} \rangle_{\mathbf{k}\omega} = \frac{T^{\alpha}}{\omega^3} i\{(\chi_{ji}^{\alpha})^{-1} - (\chi_{ij}^{\alpha*})^{-1}\}. \tag{5.56}$$

Then having used (5.54), we easily obtain the general formulas determining the spectral distributions of the electron and ion current-density fluctuations in an anisotropic nonisothermal plasma

$$\langle j_i^{\alpha} j_j^{\beta} \rangle_{\mathbf{k}\omega} = i\omega \sum_{\gamma} T^{\gamma}(\delta_{im}\delta_{\alpha\gamma} - 4\pi\chi_{ik}^{\alpha}\Lambda_{km}^{-1})^* (\delta_{in}\delta_{\beta\gamma} - 4\pi\chi_{jl}^{\beta}\Lambda_{in}^{-1})(\chi_{mn}^{\gamma*} - \chi_{nm}^{\gamma}),$$

$$\alpha = i, e; \beta = i, e. \tag{5.57}$$

Remarking that the total current $\mathbf{j}$ equals the sum of the electron and ion currents $\mathbf{j}^e$ and $\mathbf{j}^i$, we obtain the following formulas for the total current correlator and the correlators of the electron or ion currents with the total current:

$$\langle j_i j_j \rangle_{\mathbf{k}\omega} = i\omega(\delta_{im} - 4\pi\chi_{ik}\Lambda_{km}^{-1})^* (\delta_{jn} - 4\pi\chi_{jl}\Lambda_{ln}^{-1})\sum_{\gamma} T^{\gamma}(\chi_{mn}^{\gamma*} - \chi_{nm}^{\gamma}). \tag{5.58}$$

$$\langle j_i^{\alpha} j_j \rangle_{\mathbf{k}\omega} = i\omega \sum_{\gamma} T^{\gamma}(\delta_{im}\delta_{\alpha\gamma} - 4\pi\chi_{ik}^{\alpha}\Lambda_{km}^{-1})^* (\delta_{jn} - 4\pi\chi_{jl}\Lambda_{ln}^{-1})(\chi_{mn}^{\gamma*} - \chi_{nm}^{\gamma}). \tag{5.59}$$

Using the Maxwell equations and (5.57)–(5.59), it is possible to obtain the correlators of all the quantities of interest to us. In particular, the correlators of the electron-density and magnetic-field fluctuations are

$$e^2\langle(\delta n^e)^2\rangle_{\mathbf{k}\omega} = \frac{k_i k_j}{\omega^2} \langle j_i^e j_j^e \rangle_{\mathbf{k}\omega}, \tag{5.60}$$

$$e\langle\delta n^e \delta B_i\rangle_{\mathbf{k}\omega} = \frac{4\pi i}{\omega c} \frac{\eta^2}{\eta^2 - 1} \epsilon_{ikl} \frac{k_l k_j}{k^2} \langle j_j^e j_k \rangle_{\mathbf{k}\omega}, \tag{5.61}$$

$$\langle\delta B_i \delta B_j\rangle_{\mathbf{k}\omega} = \frac{16\pi^2}{\omega^2} \frac{\eta^2}{(\eta^2 - 1)^2} \epsilon_{ikl}\epsilon_{jmn} \frac{k_k k_m}{k^2} \langle j_e j_n \rangle_{\mathbf{k}\omega}. \tag{5.62}$$

As has already been noted, the correlation functions have sharp maximums near the values of ω and $\mathbf{k}$, satisfying the dispersion equation $\Lambda(\omega_s(\mathbf{k}), \mathbf{k}) = 0$ (the subscript s numbers the proper oscillations). It is easy to establish the form of the correlation functions near such maximums. For example, for the quantity $\langle j_i j_j \rangle_{\mathbf{k}\omega}$ we have

$$\langle j_i j_j \rangle_{\mathbf{k}\omega} = \sum_s B_{ij}(\omega, \mathbf{k})\, \delta(\omega - \omega_s(\mathbf{k})), \tag{5.63}$$

$$B_{ij}(\omega, \mathbf{k}) = i\pi\omega \left(\frac{\partial \Lambda}{\partial \omega}\right)^{-1} \frac{16\pi^2 \chi_{ik}^* \chi_{jl} \lambda_{km}^* \lambda_{ln} \sum_\alpha T^\alpha (\chi_{mn}^{\alpha *} - \chi_{nm}^\alpha)}{\mathrm{Im}\, \Lambda},$$

where λ_{ij} is determined from the relationship $\lambda_{ij}\Lambda_{jk} = \Lambda \delta_{ik}$.

7. Fluctuations in a Nonequilibrium Plasma

A rarefied plasma in which collisions are extremely rare may long be in a thermodynamically nonequilibrium state in which the electrons and ions are characterized by nonmaxwellian distribution functions. Fluctuations in the nonequilibrium state may be described on the basis of a generalization of the fluctuation–dissipation relationship by taking account of the smallness of the interaction between the particles and the plasma.

Let us assume that the plasma in the nonequilibrium state is homogeneous and that the nonequilibrium state is stable. If the interaction between electrons and ions is neglected completely, the correlators of the fluctuations for the electron and ion currents $\langle j_i^\alpha j_j^\alpha \rangle_{\mathbf{k}\omega}^0$ are easily found by direct calculation on the basis of a microscopic consideration. For example, in the absence of external fields these correlators are determined by (5.6), where f_0^α should be understood to be the nonequilibrium distribution function for the electrons or ions.

Let $\mathbf{E}^e$ and $\mathbf{E}^i$ denote the fields associated with the independent fluctuations of the electrons and ions. According to (5.3), these field fluctuations are connected with the fluctuating currents $\mathbf{j}^e$ and $\mathbf{j}^i$ by means of the relationship

$$j_j^\alpha = -i\omega\chi_{ij}^\alpha E_j^\alpha, \qquad \alpha = e, i, \tag{5.64}$$

where χ_{ij}^{α} is the electron or ion component of the electrical susceptibility of the nonequilibrium plasma. Solving (5.64) for $\mathbf{E}^{\alpha}$, we
may establish the following relationship between the correlators
of the fluctuating fields and the fluctuating currents:

$$\langle E_i^{\alpha} E_j^{\alpha} \rangle_{\mathbf{k}\omega} = \frac{1}{\omega^2} (\chi_{ik}^{\alpha*})^{-1} (\chi_{jl}^{\alpha})^{-1} \langle j_k^{\alpha} j_l^{\alpha} \rangle_{\mathbf{k}\omega}^0, \quad \alpha = e, i. \tag{5.65}$$

Let us note that

$$\langle E_i^{\alpha} E_j^{\beta} \rangle_{\mathbf{k}\omega} = 0, \quad \alpha \neq \beta, \tag{5.66}$$

since the electron and ion fluctuations are independent when
interaction between particles in the plasma is neglected.

Now, let us consider fluctuations in a plasma, taking account of
the interaction between electrons and ions described by means of
the self-consistent field. In this case (5.64) should be rewritten as

$$j_i^{\alpha} = -i\omega\chi_{ij}^{\alpha}(E_j + E_j^{\alpha}), \tag{5.67}$$

where the self-consistent field $\mathbf{E}$ is determined by the system of
Maxwell equations

$$\Lambda_{ij}^0 E_j = \frac{4\pi}{i\omega} \sum_{\alpha} j_i^{\alpha}. \tag{5.68}$$

Substituting (5.67) into (5.68), the system of Maxwell equations
may be reduced to the form

$$\Lambda_{ij} E_j = -4\pi \sum_{\alpha} \chi_{ij}^{\alpha} E_j^{\alpha}. \tag{5.69}$$

Solving the system (5.69) for $\mathbf{E}$ and substituting the value found
for $\mathbf{E}$ into (5.67), we obtain the following relationship between
the fluctuating current in the plasma $\mathbf{j}^{\alpha}$ and the random fluctuating
field $\mathbf{E}^{\alpha}$:

$$j_i^{\alpha} = -i\omega \sum_{\beta} (\delta_{ij}\delta_{\alpha\beta} - 4\pi\chi_{ik}^{\alpha}\Lambda_{kj}^{-1}) \chi_{jl}^{\beta} E_l^{\beta}. \tag{5.70}$$

Using this relationship and taking account of (5.65) we find:

$$\langle j_i^{\alpha} j_j^{\beta} \rangle_{\mathbf{k}\omega} = \sum_{\gamma} (\delta_{im}\delta_{\alpha\gamma} - 4\pi\chi_{ik}^{\alpha}\Lambda_{km}^{-1})^* (\delta_{jn}\delta_{\beta\gamma} - 4\pi\chi_{jl}^{\beta}\Lambda_{ln}^{-1}) \langle j_m^{\gamma} j_n^{\gamma} \rangle_{\mathbf{k}\omega}^0. \tag{5.71}$$

This formula establishes the general connection between the correlator of the current fluctuations in the plasma taking into account self-consistent interaction between the charged particles $\langle j_i^\alpha j_j^\beta \rangle_{\mathbf{k}\omega}$, and the correlator of the current fluctuations of the independent particles $\langle j_m^\gamma j_n^\gamma \rangle_{\mathbf{k}\omega}^0$, which is assumed known. In an analogous manner, it is possible to obtain formulas for the total current correlator and the correlators of the electron and ion currents with the total current:

$$\langle j_i j_j \rangle_{\mathbf{k}\omega} = (\delta_{im} - 4\pi\chi_{ik}\Lambda_{km}^{-1})^* (\delta_{jn} - 4\pi\chi_{jl}\Lambda_{ln}^{-1}) \langle j_m j_n \rangle_{\mathbf{k}\omega}^0, \tag{5.72}$$

$$\langle j_i^\alpha j_j \rangle_{\mathbf{k}\omega} = \sum_\beta (\delta_{im}\delta_{\alpha\beta} - 4\pi\chi_{ik}^\alpha\Lambda_{km}^{-1})^* (\delta_{jn} - 4\pi\chi_{jl}\Lambda_{ln}^{-1}) \langle j_m^\beta j_n \rangle_{\mathbf{k}\omega}^0. \tag{5.73}$$

Let us present a formula for the spectral distribution of the electric field fluctuations in a nonequilibrium plasma (5), which may easily be calculated by using (5.68) and (5.72):

$$\langle E_i E_j \rangle_{\mathbf{k}\omega} = \frac{16\pi^2}{\omega^2} \Lambda_{ik}^{*-1}\Lambda_{jl}^{-1} \langle j_k j_l \rangle_{\mathbf{k}\omega}^0. \tag{5.74}$$

In the equilibrium case when the distribution functions of the particles in the plasma are maxwellian, the total current correlator of the noninteracting particles $\langle j_i j_j \rangle_{\mathbf{k}\omega}^0$ is expressed in terms of the plasma dielectric permittivity tensor by means of the relation (2.56). The formulas (5.72) and (5.74) hence go over into the corresponding formulas (2.37) and (2.41). The formulas for fluctuations in a nonisothermal plasma (5.57)–(5.59) may be obtained from (5.71)–(5.73) by expressing the electron and ion current correlators in terms of the electron and ion susceptibilities of the plasma

$$\langle j_i^\alpha j_j^\alpha \rangle_{\mathbf{k}\omega}^0 = i\omega T_\alpha(\chi_{ij}^{\alpha*} - \chi_{ji}^\alpha).$$

In the general case, in the absence of thermodynamic equilibrium the correlators of the currents of the noninteracting particles $\langle j_i^\alpha j_l^\alpha \rangle_{\mathbf{k}\omega}^0$ are not expressed in terms of the electron and ion susceptibilities of the plasma χ_{ij}^α, hence, giving the susceptibilities χ_{ij}^α in a nonequilibrium plasma turns out to be inadequate to a complete description of the fluctuations, as contrasted with the case of an equilibrium or quasi-equilibrium nonisothermal plasma.

Thus, in the absence of external fields the current correlator $\langle j_i^\alpha j_j^\alpha \rangle_{\mathbf{k}\omega}^0$ in a nonequilibrium plasma is determined by the expression

$$\langle j_i^\alpha j_j^\alpha \rangle_{\mathbf{k}\omega}^0 = 2\pi e^2 n_0 \int d\mathbf{v}\, f_0^\alpha(v) v_i v_j \delta(\omega - \mathbf{kv}). \qquad (5.75)$$

The dielectric permittivity and electrical susceptibility tensors of the plasma are hence determined by the expressions

$$\epsilon_{ij} = \delta_{ij} + 4\pi \sum_\alpha \chi_{ij}^\alpha, \qquad \chi_{ij}^\alpha = \frac{e^2 n_0}{m_\alpha \omega} \int \frac{v_i(\partial f_0^\alpha / \partial v_j)}{\mathbf{kv} - \omega - io}\, dv. \qquad (5.76)$$

The formulas for fluctuations in a nonequilibrium plasma simplify in an essential way in the particular case of an isotropic distribution of the plasma particles. Introducing the longitudinal and transverse permittivities and susceptibilities for the nonequilibrium plasma ϵ_l, ϵ_t, χ_l^α, and χ_t^α, and noting that in the isotropic case

$$\Lambda_{ij}^{-1} = \frac{k_i k_j}{k^2} \epsilon_l^{-1} + \left(\delta_{ij} - \frac{k_i k_j}{k^2}\right)(\epsilon_t - \eta^2)^{-1},$$

the spectral distribution for the electron current fluctuations in a nonequilibrium plasma (5.71) may be transformed to

$$\langle j_i^e j_j^e \rangle_{\mathbf{k}\omega} = \frac{2\pi e^2 n_0 \omega}{k^2} \left\{ \int d\mathbf{v}\, f_0^e(v) \delta(\omega - \mathbf{kv}) \left[\frac{k_i k_j}{k^2} \frac{|1 + 4\pi \chi_t^i|^2}{|\epsilon_l|^2} \right.\right.$$

$$+ \frac{1}{2}\left(\delta_{ij} - \frac{k_i k_j}{k^2}\right) \frac{[\mathbf{kv}]^2}{\omega^2} \frac{|1 + 4\pi \chi_t^i \eta^2|^2}{|\epsilon_t - \eta^2|^2} \Big]$$

$$+ \int d\mathbf{v}\, f_0^i(v) \delta(\omega - \mathbf{kv}) \left[\frac{k_i k_j}{k^2} \frac{16\pi^2 |\chi_l^e|^2}{|\epsilon_l|^2} \right.$$

$$\left.\left. + \frac{1}{2}\left(\delta_{ij} - \frac{k_i k_j}{k^2}\right) \frac{[\mathbf{kv}]^2}{\omega^2} \frac{16\pi^2 |\chi_t^e|^2}{|\epsilon_t - \eta^2|^2} \right] \right\}. \qquad (5.77)$$

The spectral distribution of the electron-density fluctuations in a nonequilibrium plasma is determined by the formula

$$\langle (\delta n_e)^2 \rangle_{\mathbf{k}\omega} = \frac{2\pi n_0}{|\epsilon_l|^2} \left\{ |1 + 4\pi \chi_l^i|^2 \int d\mathbf{v}\, f_0^e(v) \delta(\omega - \mathbf{kv}) \right.$$

$$\left. + 16\pi^2 |\chi_l^e|^2 \int d\mathbf{v}\, f_0^i(v) \delta(\omega - \mathbf{kv}) \right\}. \qquad (5.78)$$

As we see, the spectral distribution of electron-density fluctuations in the nonequilibrium case is expressed directly in terms of the one-dimensional electron and ion distribution functions in the direction $\mathbf{k}$, as well as in terms of the dielectric permittivity and electron and ion susceptibilities of the plasma. The ion fluctuations in a nonequilibrium plasma are determined by formulas analogous to (5.77) and (5.78), in which the superscripts e and i should be interchanged. Let us present formulas for the spectral distributions of the total current and electric field fluctuations in a nonequilibrium isotropic plasma:

$$
\langle j_i j_j \rangle_{\mathbf{k}\omega} = \sum_\alpha \frac{2\pi e^2 n_0 \omega^2}{k^2} \int d\mathbf{v}\, f_0^\alpha(v) \delta(\omega - \mathbf{k}\mathbf{v}) \left\{ \frac{k_i k_j}{k^2} \frac{1}{|\epsilon_l|^2} \right.
$$
$$
\left. + \frac{1}{2}\left(\delta_{ij} - \frac{k_i k_j}{k^2}\right) \frac{[\mathbf{k}\mathbf{v}]^2}{\omega^2} \frac{(1 - \eta^2)^2}{|\epsilon_t - \eta^2|^2} \right\}, \tag{5.79}
$$

$$
\langle E_i E_j \rangle_{\mathbf{k}\omega} = \sum_\alpha \frac{32\pi^3 e^2 n_0}{k^2} \int d\mathbf{v}\, f_0^\alpha(v) \delta(\omega - \mathbf{k}\mathbf{v}) \left\{ \frac{k_i k_j}{k^2} \frac{1}{|\epsilon_l|^2} \right.
$$
$$
\left. + \frac{1}{2}\left(\delta_{ij} - \frac{k_i k_j}{k^2}\right) \frac{[\mathbf{k}\mathbf{v}]^2}{\omega^2} \frac{1}{|\epsilon_t - \eta^2|^2} \right\}. \tag{5.80}
$$

8. Collective Fluctuations. Effective Temperature

The spectral distributions of the correlations functions in the plasma transparency domain have sharp delta-like maximums at the frequencies ω and the vectors $\mathbf{k}$ satisfying the dispersion equation $\operatorname{Re} \Lambda(\omega, \mathbf{k}) = 0$. It is easy to establish the form of the spectral distributions of the correlation functions near such maximums.

Since $\operatorname{Im} \Lambda \ll \operatorname{Re} \Lambda$ in the plasma transparency domain, then

$$
\Lambda_{ik}^{-1*} \Lambda_{jl}^{-1} = \pi \frac{\lambda_{ik}^* \lambda_{jl}}{|\operatorname{Im} \Lambda|} \delta(\Lambda).
$$

Moreover, remarking that the relationship $\lambda_{ij} = \operatorname{Sp} \lambda \cdot e_i e_j^*$ is valid for ω and $\mathbf{k}$ satisfying the dispersion condition, the spectral distribution for the electric field fluctuations may be represented as

$$
\langle E_i E_j \rangle_{\mathbf{k}\omega} = 8\pi^2 e_i^* e_j \tilde{T} \left| \frac{\operatorname{Sp} \lambda}{\omega} \right| \delta(\Lambda), \tag{5.81}
$$

where $\tilde{T}$ is defined by the equality

$$\tilde{T} = \frac{2\pi}{\omega} \frac{\mathrm{Sp}\,\lambda}{\mathrm{Im}\,\varLambda} \langle j_i j_j \rangle_{\mathbf{k}\omega}^0 e_i e_j{}^*. \tag{5.82}$$

Using (4.14) and (3.9) in the case of an equilibrium plasma, it is easy to see that $\tilde{T}$ equals the plasma temperature T. In the general case, the quantity $\tilde{T}$ may be considered as the effective temperature characterizing the mean value of the square of the amplitude of the fluctuating electric field oscillations in the plasma. In a non-equilibrium plasma the effective temperature may take on large values. If the state of the plasma approaches the boundary of the domain of kinetic stability of the plasma, then $\mathrm{Im}\,\varLambda \to 0$, and the effective temperature hence increases without limit. Noticing that

$$\mathrm{Im}\,\varLambda = \frac{1}{2i} \mathrm{Sp}\,\lambda \cdot (\epsilon_{ij} - \epsilon_{ji}^{*}) e_i{}^* e_j , \tag{5.83}$$

the effective temperature (5.82) may then be represented as

$$\tilde{T} = \frac{4\pi}{i\omega} \frac{\langle j_i j_j \rangle_{\mathbf{k}\omega}^0 e_i e_j{}^*}{(\epsilon_{kl}^* - \epsilon_{lk}) e_k e_l{}^*} . \tag{5.84}$$

In particular, for a nonisotropic plasma the effective temperature equals

$$\tilde{T} = \frac{\sum_\alpha T_\alpha (\chi_{ij}^{\alpha*} - \chi_{ji}^\alpha) e_i e_j{}^*}{\sum_\alpha (\chi_{kl}^{\alpha*} - \chi_{lk}^\alpha) e_k e_l{}^*} . \tag{5.85}$$

The spectral distribution of the correlation function for the partial currents in the plasma transparency domain is expressed directly in terms of the correlation function for the electric field

$$\langle j_i^\alpha j_j^\beta \rangle_{\mathbf{k}\omega} = \omega^2 \chi_{ik}^{\alpha*} \chi_{jl} \langle E_k E_l \rangle_{\mathbf{k}\omega} . \tag{5.86}$$

Such a connection between the correlation functions for the partial currents and the electric field may be obtained on the basis of relationship (5.67), if the secondary fields are omitted therein. This says that the fluctuating oscillations in the domain of the maximums are characterized not only by the dispersion and polarization but also by the connections between the various quantities, exactly as for the free waves in a plasma. For example, the magnetic field, the partial current and the partial charge are connected to the electric field of the fluctuating wave by means of the relationships

$$\mathbf{B} = \eta \left[\frac{\mathbf{k}}{k} \mathbf{E} \right], \qquad j^\alpha = -i\omega\chi^\alpha E, \qquad \rho_\alpha = -i\mathbf{k}\hat{\chi}^\alpha \mathbf{E}. \qquad (5.87)$$

Hence, the correlation functions of all the quantities in the collective fluctuations domain are expressed in terms of the correlation function of the electric field.

9. Field Fluctuations in a Plasma-Beam System

As an example of a nonequilibrium plasma, let us consider a plasma through which a neutral beam of charged particles with density n_0' and temperature T' passes (6, 7). Let us limit ourselves to the analysis of fluctuations of the longitudinal electric field. According to (5.80), the spectral distribution of the longitudinal field fluctuations is determined by the expression

$$\langle E^2 \rangle_{\mathbf{k}\omega} = \frac{16\pi^2 e^2}{k^2 \, | \, \epsilon_l \, |^2} \langle \delta n^2 \rangle^0_{\mathbf{k}\omega} . \qquad (5.88)$$

Let us select the distribution functions of the electrons and ions of the plasma and beam as

$$f_0 = \left(\frac{m}{2\pi T} \right)^{3/2} \exp\left(-\frac{mv^2}{2T} \right), \quad f_0' = \left(\frac{m}{2\pi T'} \right)^{3/2} \exp\left(-\frac{m(\mathbf{v} - \mathbf{u})^2}{2T'} \right),$$

$$(5.89)$$

where $\mathbf{u}$ is the beam velocity in the plasma. In this case the longitudinal dielectric permittivity of the system is

$$\epsilon_l(\omega, \mathbf{k}) = 1 + \sum \frac{1}{a^2k^2} \{ 1 - \varphi(z) + i \sqrt{\pi} \, z \, e^{-z^2} \}$$

$$+ \sum \frac{1}{a'^2k^2} \{ 1 - \varphi(y) + i \sqrt{\pi} \, y \, e^{-y^2} \}, \qquad (5.90)$$

where

$$y = \sqrt{\frac{3}{2}} \frac{\omega - \mathbf{k}\mathbf{u}}{ks'} .$$

The summation is over the kinds of particles in the plasma and the beam. Correspondingly, the spectral distribution of the density fluctuations of the free particles is determined by the expression

$$\langle \delta n^2 \rangle^0_{\mathbf{k}\omega} = \sqrt{6\pi} \left\{ \sum \frac{n_0}{ks} e^{-z^2} - \sum \frac{n_0'}{ks'} e^{-y^2} \right\}. \tag{5.91}$$

In the plasma transparency domain ($\operatorname{Im} \epsilon \ll \operatorname{Re} \epsilon$) the expression (5.88) may be represented as

$$\langle E^2 \rangle_{\mathbf{k}\omega} = 8\pi^2 \frac{\tilde{T}}{\omega} \delta\{\epsilon_l'(\omega, \mathbf{k})\}, \tag{5.92}$$

where the effective temperature $\tilde{T}$ is

$$\tilde{T} = \frac{2\pi e^2 \omega}{k^2 \mid \epsilon_l''(\omega, \mathbf{k}) \mid} \langle \delta n^2 \rangle^0_{\mathbf{k}\omega}. \tag{5.93}$$

The frequency ω and the wave vector $\mathbf{k}$ in (5.93) are connected by means of the relationship $\epsilon_l'(\omega, \mathbf{k}) = 0$. As the state of the system approaches the boundary of the kinetic stability domain, $\epsilon_l''(\omega, \mathbf{k}) \to 0$, and the effective temperature increases unboundedly.

Neglecting the thermal motion of the ions in both the beam and the plasma at rest, and considering the condition

$$(\omega - \mathbf{k}\mathbf{u})^2 \ll k^2 s'^2$$

to be satisfied, the effective temperature of the system may be written as

$$\tilde{T} = \frac{a(z)}{\left| 1 - \dfrac{\mathbf{k}\mathbf{u}}{k\tilde{u}} \right|} T, \tag{5.94}$$

where

$$a(z) = \frac{1 + \dfrac{n_0'}{n_0} \left(\dfrac{T}{T'} \right)^{1/2} e^{z^2}}{1 + \dfrac{n_0'}{n_0} \left(\dfrac{T}{T'} \right)^{3/2} e^{z^2}}, \qquad \tilde{u} = \frac{\omega}{k} \left\{ 1 + \frac{n_0}{n_0'} \left(\frac{T'}{T} \right)^{3/2} e^{-z^2} \right\}. \tag{5.95}$$

The quantity $\tilde{u}$ plays the part of the critical beam velocity, for which the state of the plasma-beam system turns out to be unstable when it is reached.

Henceforth, let us limit ourselves to the consideration of a low density beam $n_0' \ll n_0$. In this case the influence of the beam on dispersion of the waves in the plasma may be neglected. However, the influence of the beam will be felt essentially by the effective temperature of the fluctuating oscillations.

Fluctuating Langmuir oscillations,

$$\langle E^2 \rangle_{\mathbf{k}\omega} = 4\pi^2 \tilde{T}\{\delta[\omega - (\Omega^2 + k^2 s^2)^{1/2}] + \delta[\omega + (\Omega^2 + k^2 s^2)^{1/2}]\}, \qquad (5.96)$$

are possible in the high-frequency range of the spectrum in a plasma. The effective temperature is determined by the equalities (5.87) and (5.88) for $z^2 = 1/2a^2 k^2 + \frac{3}{2}$. The critical velocity hence turns out to be of the order of the thermal electron velocity in a plasma at rest:

$$\tilde{u} = \left(\frac{\Omega^2}{k^2} + s^2\right)^{1/2}\left\{1 + \frac{n_0}{n_0'}\left(\frac{T'}{T}\right)^{3/2} \exp\left(-\frac{1}{2a^2 k^2} - \frac{3}{2}\right)\right\}. \qquad (5.97)$$

The mean energy of the fluctuating Langmuir oscillations is

$$\langle E^2 \rangle_{\mathbf{k}} = 4\pi \tilde{T}. \qquad (5.98)$$

At beam velocities close to $\tilde{u}$ this energy may significantly exceed the thermal values.

Fluctuating sonic oscillations,

$$\langle E^2 \rangle_{\mathbf{k}\omega} = 4\pi^2 a^2 k^2 \tilde{T}\{\delta(\omega - kv_s) + \delta(\omega + kv_s)\}, \qquad (5.99)$$

where

$$\tilde{T} = T\Big/\left|1 - \frac{\mathbf{k}\mathbf{u}}{k\tilde{u}}\right|, \qquad \tilde{u} = \frac{n_0}{n_0'}\left(\frac{T'}{T}\right)^{3/2} v_s, \qquad (5.100)$$

are possible in a plasma in the low-frequency range of the spectrum. In contrast to the high-frequency range, the $\tilde{u}$ in the considered case may be both greater and less than the thermal electron velocity in a plasma at rest. The mean energy of the fluctuating sonic oscillations equals

$$\langle E^2 \rangle_{\mathbf{k}} = 4\pi a^2 k^2 \tilde{T}. \qquad (5.101)$$

Let us note that formulas (5.92)–(5.95) have been obtained on the basis of linear theory and are applicable only to fluctuations with a wave vector $\mathbf{k}$ satisfying the condition $u < \tilde{u}(\mathbf{k})$. Fluctuating oscillations for which $u > \tilde{u}(\mathbf{k})$ lead to plasma instability.

REFERENCES

1. L. Tonks and I. Langmuir, *Phys. Rev.* **33**, 195 (1929).
2. G. V. Gordeev, *ZETF* **27**, 18 (1954) (Translation available from CFSTI).
3. E. E. Salpeter, *Phys. Rev.* **120**, 1528 (1960).
4. A. I. Akhiezer, I. A. Akhiezer, and A. G. Sitenko, *ZETF* **41**, 644 (1961) (*Sov. Phys.—JETP*).
5. V. P. Silin, *ZETF* **41**, 969 (1961) (*Sov. Phys.—JETP*).
6. N. Rostoker, *Nuclear Fusion* **1**, 101 (1961).
7. L. S. Bogdankevich, A. A. Rukhadze, and V. P. Silin, *Izv. VUZ, Radiofizika* **5**, 1093 (1962) (Translation available from CFSTI).

6

Electron Plasma
in a Magnetic Field

1. Plasma Dielectric Permittivity Tensor in a
Magnetic Field

Let us investigate the electromagnetic properties of a plasma in an external permanent and homogeneous magnetic field $\mathbf{B}_0$. The presence of the permanent magnetic field leads to the appearance of plasma anisotropy, with the result that the electromagnetic field in the plasma cannot possibly be separated into longitudinal and transverse components. The plasma dielectric permittivity tensor hence no longer reduces to longitudinal and transverse components, as occurs in a plasma in the absence of an external magnetic field.

In order to find the dielectric permittivity tensor of an electron plasma in a magnetic field, let us first evaluate the correlation function of the density and current for a system of noninteracting electrons in a magnetic field. Selecting the z axis in the direction of the magnetic field $\mathbf{B}_0$, the equations of the motion trajectory of an individual electron in a magnetic field may be written as

$$x(t) = x_0 + \frac{1}{\omega_B} [v_{0x} \sin \omega_B t - v_{0y}(\cos \omega_B t - 1)],$$

$$y(t) = y_0 + \frac{1}{\omega_B} [v_{0x}(\cos \omega_B t - 1) + v_{0y} \sin \omega_B t], \qquad (6.1)$$

$$z(t) = z_0 + v_{0z}t,$$

where $\omega_B = eB_0/mc$ is the electron gyrofrequency, x_0, y_0, z_0 and v_{0x}, v_{0y}, v_{0z}, the coordinates and velocity components of the electron at initial time $t = 0$.

The space-time correlation function for the electron-density fluctuations is determined by the equality

$$\langle \delta n^2 \rangle^0_{\mathbf{r}t} = \left\langle \sum_\alpha \delta\{\mathbf{r}_1 - \mathbf{r}_\alpha(t_1)\}\, \delta\{\mathbf{r}_2 - \mathbf{r}_\alpha(t_2)\} \right\rangle, \qquad (6.2)$$

where $\mathbf{r} = \mathbf{r}_2 - \mathbf{r}_1$, $t = t_2 - t_1$ and the summation is over all electrons in unit volume. Introducing the one-particle distribution function $f_0(\mathbf{v})$, the correlation function (6.2) may be rewritten as

$$\langle \delta n^2 \rangle^0_{\mathbf{r}t} = n_0 \int d\mathbf{v}\, f_0(\mathbf{v})\, \delta(\mathbf{r} - \mathbf{r}(t) + \mathbf{r}(0)\}. \qquad (6.3)$$

The spectral distribution of the correlation function of the density fluctuations is determined by the expression

$$\langle \delta n^2 \rangle^0_{\mathbf{k}\omega} = n_0 \int d\mathbf{v}\, f_0(\mathbf{v}) \int_{-\infty}^{\infty} dt\, \exp\left[-i\mathbf{k}\{\mathbf{r}(t) - \mathbf{r}(0)\} + i\omega t\right]. \qquad (6.4)$$

Utilizing the relationships

$$e^{-ia\sin\psi} = \sum_{n=-\infty}^{\infty} J_n(a)e^{-in\psi}, \qquad \int_0^{2\pi} e^{ia\sin\psi - in\psi}\, d\psi = 2\pi J_n(a),$$

where $J_n(a)$ is the Bessel function, we obtain the following formula for the spectral distribution of the correlation function of the density fluctuations:

$$\langle \delta n^2 \rangle^0_{\mathbf{k}\omega} = 2\pi n_0 \sum_{n=-\infty}^{\infty} \int d\mathbf{v}\, f_0(\mathbf{v}) I_n^2\left(\frac{k_\perp v_\perp}{\omega_B}\right) \delta(\omega - n\omega_B - k_\parallel v_\parallel) \qquad (6.5)$$

($v_\parallel$ and $v_\perp$ are the parallel and longitudinal components of the electron velocity relative to the magnetic field $\mathbf{B}_0$).

Analogously, the spectral distribution of the correlation function of the electron current fluctuations in the presence of an external magnetic field may be found in general form:

$$\langle j_i j_j \rangle^0_{\mathbf{k}\omega} = 2\pi e^2 n_0 \sum_{n=-\infty}^{\infty} \int d\mathbf{v}\, f_0(\mathbf{v}) \Pi_{ji}(n, \mathbf{v})\, \delta(\omega - n\omega_B + k_\parallel v_\parallel), \qquad (6.6)$$

where

$$\Pi_{ij}(n, \mathbf{v}) = \begin{pmatrix} \dfrac{n^2 \omega_B^2}{k_\perp^2} J_n^2 & iv_\perp \dfrac{n\omega_B}{k_\perp} J_n J_n' & v_\parallel \dfrac{n\omega_B}{k_\perp} J_n^2 \\[2ex] -iv_\perp \dfrac{n\omega_B}{k_\perp} J_n J_n' & v_\perp^2 J_n'^2 & -iv_\parallel v_\perp J_n J_n' \\[2ex] v_\parallel \dfrac{n\omega_B}{k_\perp} J_n^2 & iv_\parallel v_\perp J_n J_n' & v_\parallel^2 J_n^2 \end{pmatrix}, \qquad (6.7)$$

$$J_n = J_n(a), \qquad J_n' = \frac{\partial J_n(a)}{\partial a}, \qquad \text{and} \qquad a = \frac{k_\perp v_\perp}{\omega_B}.$$

The tensor (6.7) is written in a coordinate system in which the z axis is directed along the external magnetic field $\mathbf{B}_0$ and the x axis lies in the plane of the vectors $\mathbf{k}$ and $\mathbf{B}_0$.

Selecting the Maxwell function as the equilibrium distribution function for the electrons, by integrating (6.3) we find the following expression for the space-time correlation function of the electron-density fluctuations:

$$\langle \delta n^2 \rangle_{\mathbf{r}t}^0 = n_0 \left(\frac{m}{2\pi T} \right)^{3/2} \frac{\omega_B^2}{4 \, | \, t \, | \, \sin^2 \dfrac{\omega_B t}{2}}$$

$$\times \exp \left\{ -\frac{m}{2T} \left[\frac{\omega_B^2}{4 \sin^2 \dfrac{\omega_B t}{2}} (x^2 + y^2) + \frac{z^2}{t^2} \right] \right\}. \qquad (6.8)$$

By using the Fourier transform it is easy to obtain the spectral distribution of the correlation function

$$\langle \delta n^2 \rangle_{\mathbf{k}\omega}^0 = n_0 \int_{-\infty}^{\infty} dt$$

$$\times \exp \left[-\frac{T}{2m} \left(2 \frac{k_x^2 + k_y^2}{\omega_B^2} (1 - \cos \omega_B t) + k_z^2 t^2 \right) + i\omega t \right]. \qquad (6.9)$$

In an analogous manner it is possible to find the space-time correlation function for the electron-current fluctuations in the presence of a magnetic field:

$$\langle j_i j_j \rangle_{\mathbf{r}t}^0 = e^2 n_0 \left(\frac{m}{2\pi T} \right)^{3/2} \frac{\omega_B^2}{4 \, | \, t \, | \, \sin^2(\omega_B t/2)} p_{ij}(\mathbf{r}, t)$$

$$\times \exp \left\{ -\frac{m}{2T} \left[\frac{\omega_B^2}{4 \sin^2(\omega_B t/2)} (x^2 + y^2) + \frac{z^2}{t^2} \right] \right\}, \qquad (6.10)$$

$$p_{11} = \frac{\omega_B{}^2}{4}\left(x^2\,\text{ctn}^2\,\frac{\omega_B t}{2} - y^2\right),$$

$$p_{12} = \frac{\omega_B{}^2}{4}\left(\frac{xy}{\sin^2(\omega_B t/2)} - (x^2+y^2)\,\text{ctn}\,\frac{\omega_B t}{2}\right),$$

$$p_{13} = \frac{\omega_B}{2}\left(x\,\text{ctn}\,\frac{\omega_B t}{2} - y\right)\frac{z}{t},$$

$$p_{22} = \frac{\omega_B{}^2}{4}\left(y^2\,\text{ctn}^2\,\frac{\omega_B t}{2} - x^2\right),$$

$$p_{23} = \frac{\omega_B}{2}\left(y\,\text{ctn}\,\frac{\omega_B t}{2} + x\right)\frac{z}{t},$$

$$p_{33} = \frac{z^2}{t^2}.$$

Let us note that $p_{ji}(\mathbf{r}, t) = p_{ij}(-\mathbf{r}, -t)$. The spectral distribution function of the correlation function of the electron-current fluctuations is determined by the expression

$$\langle j_i j_j \rangle_{\mathbf{k}\omega}^0 = e^2 n_0 \frac{T}{m} \int_{-\infty}^{\infty} dt\,\pi_{ij}(t)$$

$$\times \exp\left[-\frac{T}{2m}\left\{2\frac{k_\perp{}^2}{\omega_B{}^2}(1 - \cos\omega_B t) + k_z{}^2 t^2\right\} + i\omega t\right], \quad (6.11)$$

$$\pi_{11} = \cos\omega_B t - \frac{T}{m}\frac{k_\perp{}^2}{\omega_B{}^2}\sin^2\omega_B t,$$

$$\pi_{12} = \left[-1 + \frac{T}{m}\frac{k_\perp{}^2}{\omega_B{}^2}(1 - \cos\omega_B t)\right]\sin\omega_B t,$$

$$\pi_{13} = -\frac{T}{m}\frac{k_\perp k_z}{\omega_B}t\sin\omega_B t,$$

$$\pi_{22} = \cos\omega_B t + \frac{T}{m}\frac{k_\perp{}^2}{\omega_B{}^2}(1 - \cos\omega_B t)^2, \quad (6.12)$$

$$\pi_{23} = \frac{T}{m}\frac{k_\perp k_z}{\omega_B}t(1 - \cos\omega_B t),$$

$$\pi_{33} = 1 - \frac{T}{m}k_z{}^2 t^2,$$

$$\pi_{21} = -\pi_{12}, \qquad \pi_{31} = \pi_{13}, \qquad \pi_{32} = -\pi_{23}.$$

To simplify writing of the tensor $\pi_{ij}(t)$ we have selected the coordinate system in such a manner that the vector $\mathbf{k}$ would lie in the xz plane.

According to the inversion of the fluctuation–dissipation theorem (2.69), the dielectric permittivity tensor ϵ_{ij} is determined completely by the known spectral distribution of the current fluctuations in the medium $\langle j_i j_j \rangle^0_{\mathbf{k}\omega}$. Substituting (6.11) into (2.69) and integrating with respect to ω', we obtain the following general formula for the plasma dielectric permittivity tensor in a magnetic field*:

$$\epsilon_{ij}(\omega, \mathbf{k}) = \delta_{ij} + i\,\frac{\Omega^2}{\omega} \int_0^\infty dt\, \pi_{ij}(t)$$

$$\times \exp\left[-\frac{T}{2m}\left\{2\,\frac{k_\perp^2}{\omega_B^2}(1 - \cos\omega_B t) + k_z^2 t^2\right\} + i\omega t\right] \quad (6.13)$$

[components of the tensor $\pi_{ij}(t)$ are defined by the equalities (6.12)]. Utilizing the expansion

$$e^{\beta\cos\omega_B t} = \sum_{n=-\infty}^{\infty} I_n(\beta)e^{in\omega_B t}, \quad (6.14)$$

where $I_n(\beta)$ is the modified Bessel function, the integrals with respect to t in (6.13) may be expressed in terms of the function $\varphi(z)$ we introduced in (3.26). (Integrals of the type

$$\int_0^\infty dt\, e^{i\alpha t - q^2 t^2} = \frac{i}{\alpha}\{\varphi(z) - i\sqrt{\pi z}\, e^{-z^2}\}, \quad z \equiv \frac{\alpha}{2q}$$

are encountered in the calculations of the individual components of ϵ_{ij}.) As a result, the components of the dielectric permittivity tensor of an electron plasma in a magnetic field ϵ_{ij} may be represented as[†]:

$$\epsilon_{ij}(\omega, \mathbf{k}) = \delta_{ij} - \frac{\Omega^2}{\omega^2}$$

$$\times \left\{e^{-\beta}\sum_n \frac{z_0}{z_n}\,\pi_{ij}(z_n)[\varphi(z_n) - i\sqrt{\pi}\,z_n\exp(-z_n^2)] - 2z_0^2 b_i b_j\right\}, \quad (6.15)$$

---

* Shafranov (1) obtained an analogous formula.
† Sitenko and Stepanov (2) obtained (6.15) on the basis of the kinetic equation.

where **b** is the unit vector in the direction of the external magnetic field **B$_0$** ,

$$\beta = \frac{k_\perp^2 s^2}{3\omega_B^2}, \qquad z_n = \sqrt{\frac{3}{2}} \frac{\omega - n\omega_B}{|k_\parallel| s},$$

and the components $\pi_{ij}(z_n)$ equal

$$\pi_{ij}(z_n) =$$

$$\begin{bmatrix} \dfrac{n^2}{\beta} I_n & in(I_n' - I_n) & \dfrac{k_\parallel}{|k_\parallel|}\sqrt{\dfrac{2}{\beta}}\, nz_n I_n \\[3mm] -in(I_n' - I_n) & \left(\dfrac{n^2}{\beta} + 2\beta\right) I_n - 2\beta I_n' & -i\dfrac{k_\parallel}{|k_\parallel|}\sqrt{2\beta}\, z_n(I_n' - I_n) \\[3mm] \dfrac{k_\parallel}{|k_\parallel|}\sqrt{\dfrac{2}{\beta}}\, nz_n I_n & i\dfrac{k_\parallel}{|k_\parallel|}\sqrt{2\beta}\, z_n(I_n' - I_n) & 2z_n^2 I_n \end{bmatrix}$$

$$\text{(6.16)}$$

$$\left(I_n = I_n(\beta) \qquad \text{and} \qquad I_n' = \frac{\partial}{\partial\beta} I_n(\beta)\right).$$

The expression for the dielectric permittivity tensor (6.15) may also be obtained by substituting the spectral distribution of the current fluctuations in the form (6.6) into (2.69), and by integrating over the Maxwellian velocity distribution.

In the case of a nonequilibrium plasma the dielectric permittivity tensor may be found on the basis of inverting the fluctuation-dissipation relationship. Remarking that the change in energy of an individual electron is $n\omega_B$ and $\omega - n\omega_B$, respectively, in the transverse and longitudinal directions relative to the magnetic field, we have according to (1.24) and (6.6),

$$\omega \frac{\partial}{\partial E} \langle j_i j_i \rangle^0_{\mathbf{k}\omega} = \tfrac{1}{2}\Omega^2 \sum_n \int d\mathbf{v} \left(\frac{n\omega_B}{v_\perp} \frac{\partial f_0}{\partial v_\perp} + k_\parallel \frac{\partial f_0}{\partial v_\parallel}\right)$$
$$\times \Pi_{ij}(n, \mathbf{v})\, \delta(\omega - n\omega_B - k_\parallel v_\parallel). \qquad \text{(6.17)}$$

Substituting (6.17) into (2.71), we therefore obtain the following general expression for the dielectric permittivity tensor of a plasma in a magnetic field:

$$\epsilon_{ij}(\omega, \mathbf{k}) = \left(1 - \frac{\Omega^2}{\omega^2}\right)\delta_{ij} - \frac{\Omega^2}{\omega^2} \sum_n \int d\mathbf{v} \left(\frac{n\omega_B}{v_\perp}\frac{\partial f_0}{\partial v_\perp} + k_\parallel \frac{\partial f_0}{\partial v_\parallel}\right)$$

$$\times \frac{\Pi_{ij}(n, \mathbf{v})}{k_\parallel v_\parallel + n\omega_B - \omega - io}. \qquad \text{(6.18)}$$

The components of the tensor $\Pi_{ij}(n, \mathbf{v})$ are determined by the matrix (6.7).

According to (2.10), the electric field of electromagnetic waves in a plasma is determined by the equations

$$\left\{ \eta^2 \left(\frac{k_i k_j}{k^2} - \delta_{ij} \right) + \epsilon_{ij} \right\} E_j = 0 \qquad (6.19)$$

($\eta = kc/\omega$ is the refractive index of waves with frequency ω). The system of Eqs. (6.19) has a nonzero solution if its determinant equals zero. This is the dispersion equation connecting the frequency ω to the wave vector $\mathbf{k}$ of the electromagnetic waves in the plasma, and it may be represented as

$$A\eta^4 + B\eta^2 + C = 0, \qquad (6.20)$$

$$A = \epsilon_{11} \sin^2 \vartheta + \epsilon_{33} \cos^2 \vartheta + 2\epsilon_{13} \sin \vartheta \cos \vartheta,$$

$$B = 2(\epsilon_{12}\epsilon_{23} - \epsilon_{22}\epsilon_{13}) \sin \vartheta \cos \vartheta - (\epsilon_{11}\epsilon_{22} + \epsilon_{12}^2) \sin^2 \vartheta$$

$$- (\epsilon_{22}\epsilon_{33} + \epsilon_{23}^2) \cos^2 \vartheta - \epsilon_{11}\epsilon_{33} + \epsilon_{13}^2 , \qquad (6.21)$$

$$C = \epsilon_{11}\epsilon_{22}\epsilon_{33} + \epsilon_{11}\epsilon_{23}^2 - \epsilon_{22}\epsilon_{13}^2 + \epsilon_{33}\epsilon_{12}^2 + 2\epsilon_{12}\epsilon_{23}\epsilon_{31} .$$

ϑ is the angle between the propagation direction of the wave $\mathbf{k}$ and the magnetic field $\mathbf{B}_0$.

Let us note that the components of the dielectric permittivity tensor ϵ_{ij} are complex, which indicates the presence of absorption in the plasma. Absorption is particularly large in cyclotron resonance domains, i.e., at frequencies ω which are multiples of ω_B . Indeed, if $\omega \cong n\omega_B$, then $z_n \cong 0$ and the exponential factor $\exp(-z_n^2)$ in the imaginary part of the components of ϵ_{ij} , become on the order of unity.

Absorption in a plasma in the absence of collisions is Cerenkov in nature. The presence of a magnetic field causes an increase in the refractive index of the plasma in definite frequency ranges. Because of the significant slowing down of the waves the majority of plasma electrons turns out to possess ultralight velocities, which also leads to strong absorption of waves as a result of a reverse Cerenkov effect. In the general case the dispersion equation (6.20) is very complicated, therefore, we will limit ourselves henceforth to the analysis of the most interesting case of low temperatures (strong magnetic field) $ks \ll \omega_B$.

2. Electromagnetic Waves in a Plasma in a Magnetic Field (without Taking Account of the Thermal Electron Motion)

If thermal electron motion is neglected completely, the dielectric permittivity tensor (6.15) becomes

$$\epsilon_{ij} = \begin{pmatrix} \epsilon_1 & -i\epsilon_2 & 0 \\ i\epsilon_2 & \epsilon_1 & 0 \\ 0 & 0 & \epsilon_3 \end{pmatrix}, \qquad (6.22)$$

where

$$\epsilon_1 = 1 - \frac{\Omega^2}{\omega^2 - \omega_B{}^2}, \qquad \epsilon_2 = \frac{\omega_B}{\omega}\frac{\Omega^2}{\omega^2 - \omega_B{}^2}, \qquad \epsilon_3 = 1 - \frac{\Omega^2}{\omega^2}. \quad (6.23)$$

As we see, neglecting the thermal electron motion in the plasma leads to absence of spatial dispersion. Hence (6.22) may be considered also as the limiting value of the plasma dielectric permittivity tensor (6.15) in the large wavelength range ($k \to 0$). According to (6.22), a plasma in a magnetic field is an anisotropic and gyrotropic medium even in the absence of spatial dispersion.[*]

Neglecting the thermal electron motion, the dispersion equation is

$$A_0\eta^4 + B_0\eta^2 + C_0 = 0, \qquad (6.24)$$

$$A_0 = \epsilon_1 \sin^2 \vartheta + \epsilon_3 \cos^2 \vartheta,$$
$$B_0 = - [(\epsilon_1{}^2 - \epsilon_2{}^2) \sin^2 \vartheta + \epsilon_1\epsilon_3(1 + \cos^2 \vartheta)],$$
$$C_0 = \epsilon_3(\epsilon_1{}^2 - \epsilon_2{}^2),$$

where the coefficients A_0, B_0, and C_0 turn out to be independent of η because of the absence of spatial dispersion. Hence (6.24) may be solved directly for η, thereby determining the dependence of the refractive index of electromagnetic waves in a plasma on the frequency and direction of wave propagation. Equation (6.24) has two different solutions

$$\eta_\pm{}^2 = \frac{(\epsilon_1{}^2 - \epsilon_2{}^2) \sin^2 \vartheta + \epsilon_1\epsilon_3(1 + \cos^2 \vartheta) \pm ((\epsilon_1{}^2 - \epsilon_2{}^2 - \epsilon_1\epsilon_3)^2 \sin^4 \vartheta + 4\epsilon_2{}^2\epsilon_3{}^2 \cos^2 \vartheta)^{1/2}}{2(\epsilon_1 \sin^2 \vartheta + \epsilon_3 \cos^2 \vartheta)}, \quad (6.25)$$

[*] Ginzburg (3) considered electromagnetic waves in a plasma without taking account of spatial dispersion.

determining the refractive indices of the ordinary (η_+^2) and extraordinary (η_-^2) electromagnetic waves in the plasma.

In the general case, for an arbitrary (relative to the magnetic field) direction of propagation, the electromagnetic waves in the plasma are not transverse, and are characterized by elliptic polarization. Using (6.19) and (6.22) the complex polarization vectors for the electromagnetic waves may be selected for a given frequency as

$$e^\pm = \left\{ 1; \quad \frac{i\epsilon_2}{\eta_\pm^2 - \epsilon_1}; \quad \frac{\eta_\pm^2 \sin\vartheta \cos\vartheta}{\eta_\pm^2 \sin^2\vartheta - \epsilon_3} \right\}. \tag{6.26}$$

Let us note that the polarization vectors for the ordinary and extraordinary waves satisfy the following orthogonality condition:

$$\epsilon_{ij}e_j^+ e_i^{-*} = 0.$$

For waves being propagated along the magnetic field $(\vartheta = 0)$, the refractive indices equal

$$\eta_\pm^2 = \epsilon_1 \pm \epsilon_2,$$

and, correspondingly, the polarization vectors equal

$$e^\pm = (1; \quad \pm i; \quad 0).$$

For longitudinal propagation the electromagnetic waves are transverse, whereupon the ordinary wave possesses right circular polarization, and the extraordinary wave, left circular polarization.

For waves being propagated perpendicularly to the magnetic field direction $(\vartheta = \pi/2)$, we have

$$\eta_+^2 = \epsilon_1 - \frac{\epsilon_2^2}{\epsilon_1}, \qquad \eta_-^2 = \epsilon_3;$$

$$e^+ = \left(1; \quad -i\frac{\epsilon_1}{\epsilon_2}; \quad 0 \right), \qquad e^- = (0; \quad 0; \quad 1).$$

Neglecting the thermal electron motion, the tensor (6.22) is Hermitian, and there is no wave damping. Taking account of the anti-Hermitian portion of the tensor ϵ_{ij} by using (2.14), the imaginary part of the refractive index determining the wave damping in the plasma

$$\eta'' = \tfrac{1}{2}\eta_\pm \arg(e^* \hat{\epsilon} e) \tag{6.27}$$

may easily be found.

Strictly speaking, electromagnetic waves in a plasma in the presence of a magnetic field do not separate into longitudinal and transverse. However, if $A \equiv k_i \epsilon_{ij} k_j / k^2 \to 0$, the longitudinal component of the electric field will be considerably greater than the transverse component. This is easily seen by multiplying (6.19) by $\mathbf{k}$:

$$E_{\parallel} = -\frac{k_i \epsilon_{ij} E_{\perp j}}{kA}.$$

Hence, it is seen that for purely longitudinal oscillations in a plasma ($E_{\perp} = 0$) two conditions $A = 0$ and $A = 0$ should be satisfied. In the general case these conditions are not satisfied jointly. However, in the frequency range in which η^2 is very large, only the term with the highest power of η^2 should be retained in the dispersion equation (6.20), and hence the dispersion equation will reduce to the condition $A = 0$.

Neglecting the thermal electron motion, the dispersion equation for longitudinal waves in a plasma is

$$\epsilon_1 \sin^2 \vartheta + \epsilon_3 \cos^2 \vartheta = 0. \tag{6.28}$$

Using (6.23) for ϵ_1 and ϵ_3 we find the eigenfrequencies of the longitudinal waves in the plasma from (6.28):

$$\omega_{\pm}^2 = \tfrac{1}{2}(\Omega^2 + \omega_B^2) \pm \tfrac{1}{2}[(\Omega^2 + \omega_B^2)^2 - 4\Omega^2 \omega_B^2 \cos^2 \vartheta]^{1/2}. \tag{6.29}$$

Proper oscillations of the plasma in a magnetic field correspond to these frequencies. In the limiting case $B_0 = 0$ we easily obtain an expression for the eigenfrequency of the longitudinal oscillations in an isotropic plasma from (6.29).

3. Electromagnetic Waves in a Plasma in a Magnetic Field (Taking Account of Thermal Electron Motion)

Now, let us investigate the effects in a magnetoactive plasma, which are associated with taking account of the thermal electron motion. Let us limit ourselves to an analysis of the low-temperature (strong magnetic field) case for which the condition $\beta \ll 1$ is satisfied. Far from the resonance frequencies $|z_n| \gg 1$ the asymptotic expression (3.28) may be used for the function $\varphi(z_n)$. Expanding the functions $e^{-\beta}$ and $I_n(\beta)$ in power series in β, we obtain the following expressions for the ϵ_{ij} components:

$$\epsilon_{11} = 1 - \frac{v}{1-u} - a^2 k^2 v^2 \left[\frac{1+3u}{(1-u)^3} \cos^2 \vartheta + \frac{3}{(1-u)(1-4u)} \sin^2 \vartheta \right],$$

$$\epsilon_{12} = -i \sqrt{u} \, \frac{v}{1-u} - ia^2 k^2 \sqrt{u} v^2 \left[\frac{3+u}{(1-u)^3} \cos^2 \vartheta \right.$$

$$\left. + \frac{3}{(1-u)(1-4u)} \sin^2 \vartheta \right],$$

$$\epsilon_{13} = -2a^2 k^2 \, \frac{v^2}{(1-u)^2} \sin \vartheta \cos \vartheta,$$

$$(6.30)$$

$$\epsilon_{22} = 1 - \frac{v}{1-u} - a^2 k^2 v^2 \left[\frac{1+3u}{(1-u)^3} \cos^2 \vartheta + \frac{1+8u}{(1-u)(1-4u)} \sin^2 \vartheta \right],$$

$$\epsilon_{23} = ia^2 k^2 \sqrt{u} v^2 \, \frac{3-u}{(1-u)^2} \sin \vartheta \cos \vartheta,$$

$$\epsilon_{33} = 1 - v - a^2 k^2 v^2 \left[3 \cos^2 \vartheta + \frac{1}{1-u} \sin^2 \vartheta \right],$$

where we have introduced the notation $u = \omega_B^2/\omega^2$ and $v = \Omega^2/\omega^2$. The members proportional to $a^2 k^2$ in (6.30) take account of the thermal electron motion in the plasma, which also specifies the spatial dispersion of the plasma.

Using (6.30), let us evaluate the coefficients A, B, and C, which are in the dispersion equation (6.20):

$$A = A_0 + A_1 \eta^2, \quad B = B_0 + B_1 \eta^2, \quad C = C_0 + C_1 \eta^2,$$

$$A_1 = -\frac{1}{3} \frac{s^2}{c^2} v \left\{ 3 \cos^2 \vartheta + \frac{6-3u+u^2}{(1-u)^3} \sin^2 \vartheta \cos^2 \vartheta \right.$$

$$\left. + \frac{3}{(1-u)(1-4u)} \sin^4 \vartheta \right\},$$

$$B_1 = \frac{1}{3}\frac{s^2}{c^2}\, v \left\{ \frac{2(1 + u - v)}{(1 - u)^2} \sin^2 \vartheta \cos^2 \vartheta \right.$$

$$+ \frac{1 + \cos^2 \vartheta}{1 - u}\left[(1 - u - v)\left(3\cos^2\vartheta + \frac{\sin^2\vartheta}{1-u}\right)\right.$$

$$\left.+ (1 - v)\left(\frac{1 + 3u}{(1-u)^2}\cos^2\vartheta + \frac{3\sin^2\vartheta}{1 - 4u}\right)\right]$$

$$+ \frac{2\sin^2\vartheta}{(1-u)^2}\left[\frac{1 + 3u - v - uv}{1 - u}\cos^2\vartheta\right.$$

$$\left.\left.+ \frac{2(1-u)(1 + 2u - v)}{1 - 4u}\sin^2\vartheta\right]\right\},$$

$$C_1 = -\frac{1}{3}\frac{s^2}{c^2}\, v \left\{ \frac{2(1 - v)}{(1 - u)^2}\left[\frac{1 + 3u - v - uv}{1 - u}\cos^2\vartheta\right.\right.$$

$$\left.+ \frac{2(1-u)(1 + 2u - v)}{1 - 4u}\sin^2\vartheta\right]$$

$$\left.+ \frac{(1 - v)^2 - u}{1 - u}\left(3\cos^2\vartheta - \frac{\sin^2\vartheta}{1 - u}\right)\right\}.$$

The quantities A_1, B_1, and C_1 are proportional to the square of the ratio between the mean thermal electron velocity and the velocity of light, and they characterize corrections to the zero approximation (6.24).

Hence, the dispersion equation (6.20) becomes

$$A_1\eta^6 + (A_0 + B_1)\eta^4 + (B_0 + C_1)\eta^2 + C_0 = 0. \qquad (6.31)$$

The three roots of this equation $\eta_1{}^2$, $\eta_2{}^2$, and $\eta_3{}^2$, respectively, determine the refractive indices of the ordinary, extraordinary, and plasma waves. The refractive indices of the ordinary and extraordinary waves equal

$$\eta_{1,2}^2 = (1 + \epsilon_\pm)\eta_\pm{}^2,$$

$$\epsilon_\pm = -(A_1\eta_\pm{}^4 + B_1\eta_\pm{}^2 + C_1)/(2A_0\eta_\pm{}^2 + B_0), \qquad |\epsilon_\pm| \ll 1. \qquad (6.32)$$

Since $s^2 \ll c^2$ under ordinary conditions, the thermal corrections to the refractive indices $\eta_{1,2}^2$ are very slight.

If A_0 is small in absolute value as compared to unity, i.e., ω^2 is close to ω_+^2 or ω_-^2, we then obtain the approximate expressions

$$\eta_1^2 = -\frac{C_0}{B_0}; \qquad \eta_2^2 = -\frac{B_0}{A_0} \qquad (6.33)$$

from (6.24). Since $B_0(\omega) > 0$ for $\omega^2 \simeq \omega_\pm^2$, then $\eta_2^2 \to +\infty$ as $\omega^2 \to \omega_+^2$ (or ω_-^2) from $\omega^2 < \omega_+^2$ (or $\omega^2 < \omega_-^2$); and $\eta_2^2 \to -\infty$ for $\omega^2 \to \omega_+^2$ (or ω_-^2) from $\omega^2 > \omega_+^2$ (or $\omega^2 > \omega_-^2$). However, let us note that for very small A_0 the expression (6.33) for η_2^2 has no meaning since it has been obtained under the condition

$$\frac{s^2}{c^2} \ll |A_0(\omega)| \ll 1. \qquad (6.34)$$

Upon compliance with condition (6.34) we obtain for the refractive index of the plasma wave

$$\eta_3^2 = -\frac{A_0}{A_1}. \qquad (6.35)$$

If the thermal electron motion is neglected, then A_1 vanishes. Since the refractive index for the plasma waves is finite, it is hence necessary that A_0 also vanish. This condition is indeed the dispersion equation for plasma waves neglecting the thermal electron motion (6.28).

Formulas (6.30) are applicable far from the resonance frequencies. In the resonant frequencies range $\omega \simeq n\omega_B$, strong absorption of the electromagnetic waves occurs, which is due to the thermal electron motion. As an example, let us consider the case of frequencies close to the cyclotron frequency $\omega \simeq \omega_B$.

Putting $\beta \ll 1$ and $|z_1| \ll 1$, we obtain from (6.15)

$$\epsilon_{11} = \epsilon_{22} = 1 - \frac{1}{4}v + i\sqrt{\frac{3\pi}{8}}\frac{v}{(\eta s/c)\cos\vartheta}.$$

$$\epsilon_{12} = \frac{i}{4}v - \sqrt{\frac{3\pi}{8}}\frac{v}{(\eta s/c)\cos\vartheta}, \qquad (6.36)$$

$$\epsilon_{33} = 1 - v, \qquad \epsilon_{13} = \epsilon_{23} = 0.$$

Substituting these expressions in the dispersion equation (6.20) and retaining terms proportional to $c/s \gg 1$, we find

$$\eta_{1,2}^2 = (1 + \delta_\pm)\eta_\pm^2, \tag{6.37}$$

$$\delta_\pm = i\sqrt{\frac{8}{3\pi}}\frac{s}{c}\frac{\cos\vartheta}{v\eta_\pm}\{[1 - v(\tfrac{1}{4}\sin^2\vartheta + \cos^2\vartheta)]\eta_\pm^4$$
$$-[(1 - v)(1 - \tfrac{1}{4}v)(1 + \cos^2\vartheta) + (1 - \tfrac{1}{2}v)\sin^2\vartheta]\eta_\pm^2$$
$$+ (1 - v)(1 - \tfrac{1}{2}v)\}[2\eta_\pm^2\sin^2\vartheta + 2(v - 1) - \sin^2\vartheta]^{-1}.$$

For $\omega \simeq \omega_B$ the damping coefficient equals s/c in order of magnitude, i.e., is considerably greater than the customary thermal corrections to the refractive indices of the ordinary and extra-ordinary waves, which are proportional to s^2/c^2.

4. Longitudinal Plasma Oscillations in a Magnetic Field

Let us examine in more detail the problem of the longitudinal plasma oscillations. In the limiting case $\eta^2 \gg 1$, a longitudinal plasma wave may be isolated from (6.20), for which the dispersion equation will be

$$A(\omega, \mathbf{k}) \equiv \epsilon_{11}(\omega, \mathbf{k})\sin^2\vartheta + \epsilon_{33}(\omega, \mathbf{k})\cos^2\vartheta + 2\epsilon_{13}(\omega, \mathbf{k})\sin\vartheta\cos\vartheta = 0. \tag{6.38}$$

Henceforth, we shall consider the wave vector $\mathbf{k}$ to be real, hence (6.38) will define the complex frequency of the wave $\omega \to \omega' = \omega - i\gamma$ (γ the damping coefficient) as a function of the quantity $\mathbf{k}$. Using (6.15), we may rewrite (6.38) in the following manner:

$$1 + a^2k^2 - e^{-\beta}\sum_{n=-\infty}^{\infty}\frac{z_0}{z_n}I_n(\beta)[\varphi(z_n) - i\sqrt{\pi}z_n\,e^{-z_n^2}] = 0, \tag{6.39}$$

where

$$\beta = \frac{k_\perp^2 s^2}{3\omega_B^2} \quad \text{and} \quad z_n = \sqrt{\frac{3}{2}}\frac{\omega' - n\omega_B}{|k_z|s}.$$

If $k_\perp = 0$, then $\beta = 0$ and the dispersion equation (6.39) has the same form as when the magnetic field is absent; therefore, the magnetic field does not affect plasma waves being propagated along the field.

For an arbitrary direction of wave propagation in the low plasma temperature case $(\beta \ll 1)$, by expanding the functions $e^{-\beta}$ and $I_n(\beta)$ in powers of β and using the asymptotic expansion (3.28) for the function $\varphi(z_n)$, we obtain

$$a^2 k^2 - \left(\frac{1}{2z_0^2} + \frac{3}{4z_0^4} + \cdots \right)$$

$$- \beta \left[\frac{z_0}{2} \left(\frac{1}{z_1} + \frac{1}{z_{-1}} + \frac{1}{2z_1^3} + \frac{1}{2z_{-1}^3} \right) - 1 - \frac{1}{2z_0^2} \right]$$

$$- \beta^2 \left[\frac{z_0}{8} \left(\frac{1}{z_2} + \frac{1}{z_{-2}} - \frac{4}{z_1} - \frac{4}{z_{-1}} \right) + \frac{3}{4} \right] + \cdots$$

$$+ i \sqrt{\pi} z_0 \left\{ (1 - \beta + \tfrac{3}{4}\beta^2) \exp(-z_0^2) \right.$$

$$+ \frac{\beta}{2} (1 - \beta)[\exp(-z_1^2) + \exp(-z_{-1}^2)]$$

$$\left. + \frac{\beta^2}{8} [\exp(-z_2^2) + \exp(-z_{-2}^2)] + \cdots \right\} = 0. \qquad (6.40)$$

If the thermal electron motion is neglected completely $(\beta = 0)$, then (6.40) reduces to the zero approximation dispersion equation (6.28). Taking into account that $a^2 k^2 \ll 1$, we look for the solutions of (6.40) in the form

$$\omega_{1,2}^2 = (1 + \epsilon_\pm)\omega_\pm^2, \qquad |\epsilon_\pm| \ll 1. \qquad (6.41)$$

We obtain for the corrections $\epsilon_\pm$ to the proper frequencies $\omega_{1,2}$

$$\epsilon_\pm = 3a^2 k^2 \frac{v_\pm^3}{1 + v_\pm u_\pm (1 - u_\pm)^{-2} \sin^2 \vartheta}$$

$$\times \left\{ \cos^4 \vartheta + \frac{2 - u_\pm + \frac{1}{3} u_\pm^2}{(1 - u_\pm)^3} \sin^2 \vartheta \cos^2 \vartheta + \frac{\sin^4 \vartheta}{(1 - u_\pm)(1 - 4u_\pm)} \right\}, \qquad (6.42)$$

$$v_\pm = \frac{\Omega^2}{\omega_\pm^2}, \qquad u_\pm = \frac{\omega_B^2}{\omega_\pm^2}.$$

Hence, in the case of low plasma temperatures (strong magnetic fields) $ks \ll \omega_B$, two proper frequencies of the plasma oscillations exist which are determined by (6.41) and (6.42). We find the damping corresponding to these frequencies by taking account of exponentially small terms in (6.40):

$$\gamma_{1,2} = \sqrt{\frac{\pi}{8}} \frac{\omega_\pm^2}{a^3 k^3 \Omega \cos \vartheta} \frac{1}{1 + \omega_B^2 \Omega^2 \sin^2 \vartheta/(\omega_\pm^2 - \omega_B^2)^2}$$

$$\times \left\{ \exp \left(-\frac{\omega_{1,2}^2}{2a^2 k^2 \Omega^2 \cos^2 \vartheta} \right) \right.$$

$$+ \tfrac{1}{2} a^2 k^2 \frac{\Omega^2}{\omega_B^2} \sin^2 \vartheta \left[\exp \left(-\frac{(\omega_{1,2} - \omega_B)^2}{2a^2 k^2 \Omega^2 \cos^2 \vartheta} \right) \right.$$

$$\left. \left. + \exp \left(-\frac{(\omega_{1,2} + \omega_B)^2}{2a^2 k^2 \Omega^2 \cos^2 \vartheta} \right) \right] + \cdots \right\}. \tag{6.43}$$

The expression (6.41) has been obtained under the condition $| z_n | \gg 1$ ($| \epsilon_\pm | \ll 1$). If $\vartheta \to 0$, then $\omega_1 \simeq \omega_+ \to \omega_B$ (for $\Omega < \omega_B$) and $\omega_2 \simeq \omega_- \to \omega_B$ (for $\Omega > \omega_B$). In this case the inequality $| z_1 | \gg 1$ is not satisfied, and (6.41) have no meaning. We find from the condition $| z_1 | \gg 1$ (or from the condition $| \epsilon_\pm | \ll 1$) that the applicability of (6.41) for ω_1 when $\Omega < \omega_B$, and for ω_2 when $\Omega > \omega_B$, is limited by the condition

$$\vartheta^2 \gg ak \frac{| \Omega^2 - \omega_B^2 |}{\omega_B \Omega}.$$

As follows from the exact dispersion equation (6.39), a natural solution $\omega \simeq \Omega$ exists for $\vartheta = 0$ and $ak \ll 1$.

If $\Omega > \omega_B$, then for some ω_B and Ω there exists an angle $\vartheta = \vartheta_n$, for which the frequency ω_1, defined by (6.41), turns out to be a multiple of ω_B:

$$\omega_1 \simeq n\omega_B \qquad (n = 2, 3, \ldots). \tag{6.44}$$

However, the dispersion equation (6.40) itself has been obtained under the assumption $| z_n | \gg 1$. Hence, it is impossible to use (6.41) for angles close to ϑ_n, if (6.44) holds for these angles. In order to obtain a dispersion equation valid for $\vartheta \simeq \vartheta_n$, the function $\varphi(z_n)$ should be retained in (6.39), and the asymptotic expansion (3.28) should be used for the rest, as before:

$$a^2 k^2 - \frac{1}{2z_0^2} - \beta \left[\frac{z_0}{2} \left(\frac{1}{z_1} + \frac{1}{z_{-1}} \right) - 1 \right] + \cdots$$

$$- \frac{\beta^n}{n! \, 2^n} \frac{z_0}{z_n} [\varphi(z_n) - i \sqrt{\pi} z_n \, e^{-z_n^2}] = 0. \qquad (6.45)$$

If ϑ_n is not close to $\pi/2$, then by putting $\varphi(z_n) \simeq 0$, we find

$$\omega' = n\omega_B - i\gamma_n , \qquad (6.46)$$

where

$$\gamma_n = \frac{\sqrt{\pi} n^4 \sin^{2n} \vartheta}{2^{n+3/2} 3^{n-3/2} n! \cos^3 \vartheta (1 + n^4 (n^2 - 1)^{-2} \tan^2 \vartheta)}$$

$$\times \left(\frac{ks}{\omega_B} \right)^{2n-4} \cdot ks, \qquad n = 2, 3, \dots . \qquad (6.47)$$

Waves with frequencies which are multiples of ω_B damp strongly for ϑ not close to $\pi/2$. The damping coefficient γ_n is proportional to $(ks/\omega_B)^{2n-4}$ and diminishes as n increases. For $n = 2$ the damping is an order of magnitude $(ak)^{-1}$ greater than the customary thermal corrections to the frequency. If $\vartheta_n \simeq \pi/2$, then the function $\varphi(z_n) = 1$ for $z_n \gg 1$ and, therefore, we obtain

$$\omega = n\omega_B \pm \Delta_n , \qquad \Delta_n = \frac{n^2 - 1}{(2^{n+1} 3^{n-1} n!)^{1/2}} \left(\frac{ks}{\omega_B} \right)^{n-2} ks. \qquad (6.48)$$

Hence, for $\vartheta = \pi/2$ and a given magnetic field longitudinal waves may not be propagated in a plasma in the frequency band $n\omega_B - \Delta_n < \omega < n\omega_B + \Delta_n$. The slit width $2\Delta_n$ diminishes as n increases. [Gross (4) indicated the existence of a slit in the plasma oscillation spectrum for $n = 2$.] Finally, if $\omega_B \simeq \Omega$, then (6.41) are not applicable for small angles ϑ, since the condition $| z_1 | \gg 1$ is not satisfied here. In this case the exact dispersion equation (6.39) becomes

$$1 - \frac{\Omega^2}{\omega^2} - \frac{\vartheta^2}{2} \frac{z_0}{z_1} [\varphi(z_1) - i \sqrt{\pi} z_1 \exp(-z_1^2)] = 0. \qquad (6.49)$$

Assuming $|z_1| \ll 1$, we find

$$\omega = \Omega \simeq \omega_B, \qquad \gamma = \frac{1}{4}\sqrt{\frac{\pi}{2}}\frac{\vartheta^2}{ak}\Omega, \qquad \vartheta^2 \ll ak, \qquad (6.50)$$

i.e., for $\omega_B \simeq \Omega$ the plasma wave with frequency $\omega = \Omega \simeq \omega_B$ damps strongly.

In conclusion, let us present formulas for the frequency and damping of oscillations of a plasma in a weak magnetic field $(\omega_B \ll \Omega)$:

$$\omega^2 = \Omega^2 + k^2 s^2 + \omega_B{}^2 \sin^2 \vartheta, \qquad (6.51)$$

$$\gamma = \sqrt{\frac{\pi}{8}}\frac{\Omega}{a^3 k^3}\left(1 + \frac{\omega_B{}^2 \sin^2 \vartheta}{4!\,(ak)^6\,\Omega^2}\right)\exp\left[-\frac{\omega^2}{2a^2 k^2\,\Omega^2}\right]. \qquad (6.52)$$

5. Electromagnetic Fluctuations in an Electron Plasma in the Presence of a Magnetic Field

Fluctuations in an electron plasma in a permanent, homogeneous magnetic field $\mathbf{B}_0$ are determined by the general formulas of Chapter 2, obtained for an anisotropic medium. According to (2.37), the spectral distribution of the current density fluctuations in a magnetoactive plasma is determined for $T \gg \hbar\omega$ by

$$\langle j_i j_j \rangle_{\mathbf{k}\omega} = i\frac{T}{\omega}\{\alpha_{ij}^*(\omega, \mathbf{k}) - \alpha_{ji}(\omega, \mathbf{k})\}, \qquad (6.53)$$

where

$$\alpha_{ij}(\omega, \mathbf{k}) = \frac{\omega^2}{4\pi}\{\Lambda_{ij}^0 - \Lambda_{ik}^0 \Lambda_{kl}^{-1}\Lambda_{lj}^0\}. \qquad (6.54)$$

The components of the tensor Λ_{ij} for a plasma in a magnetic field are

$$\Lambda_{ij} \equiv \begin{pmatrix} -\eta^2 \cos^2 \vartheta + \epsilon_{11} & \epsilon_{12} & \eta^2 \sin \vartheta \cos \vartheta + \epsilon_{13} \\ -\epsilon_{12} & -\eta^2 + \epsilon_{22} & \epsilon_{23} \\ \eta^2 \sin \vartheta \cos \vartheta + \epsilon_{13} & -\epsilon_{23} & -\eta^2 \sin^2 \vartheta + \epsilon_{33} \end{pmatrix} \qquad (6.55)$$

(the z axis is directed along $\mathbf{B}_0$, the x axis lies in the plane of the vectors $\mathbf{k}$ and $\mathbf{B}_0$, ϑ is the angle between $\mathbf{k}$ and $\mathbf{B}_0$).

The inverse tensor Λ_{ij}^{-1} is defined by the equality

$$\Lambda_{ij}^{-1} = \frac{\lambda_{ij}}{\Lambda},\tag{6.56}$$

where λ_{ij} is a cofactor and Λ the determinant of the matrix (6.55). The determinant Λ equals

$$\Lambda(\omega, \mathbf{k}) \equiv A\eta^4 + B\eta^2 + C,\tag{6.57}$$

where the coefficients A, B, and C are expressed in terms of the components of the tensor ϵ_{ij} according to (6.21). The cofactors λ_{ij} are defined by the matrices

$$\lambda_{ij} \equiv \begin{pmatrix}
(\eta^2 - \epsilon_{22})(\eta^2 \sin^2 \vartheta - \epsilon_{33}) + \epsilon_{23}^2 \\
(\eta^2 \sin \vartheta \cos \vartheta + \epsilon_{13}) \epsilon_{23} - (\eta^2 \sin^2 \vartheta - \epsilon_{33}) \epsilon_{12} \\
(\eta^2 - \epsilon_{22})(\eta^2 \sin \vartheta \cos \vartheta + \epsilon_{13}) + \epsilon_{12}\epsilon_{23}
\end{pmatrix}$$

$$\begin{pmatrix}
(\eta^2 \sin^2 \vartheta - \epsilon_{33}) \epsilon_{12} - (\eta^2 \sin \vartheta \cos \vartheta + \epsilon_{13}) \epsilon_{23} \\
(\eta^2 \cos^2 \vartheta - \epsilon_{11})(\eta^2 \sin^2 \vartheta - \epsilon_{33}) - (\eta^2 \sin \vartheta \cos \vartheta + \epsilon_{13})^2 \\
(\eta^2 \sin \vartheta \cos \vartheta + \epsilon_{13}) \epsilon_{12} - (\eta^2 \cos^2 \vartheta - \epsilon_{11}) \epsilon_{23}
\end{pmatrix}$$

$$\begin{pmatrix}
(\eta^2 - \epsilon_{22})(\eta^2 \sin \vartheta \cos \vartheta + \epsilon_{13}) + \epsilon_{12}\epsilon_{23} \\
(\eta^2 \cos^2 \vartheta - \epsilon_{11}) \epsilon_{23} - (\eta^2 \sin \vartheta \cos \vartheta + \epsilon_{13}) \epsilon_{12} \\
(\eta^2 \cos^2 \vartheta - \epsilon_{11})(\eta^2 - \epsilon_{22}) + \epsilon_{12}^2
\end{pmatrix}.\tag{6.58}$$

Fluctuations in the plasma density in a magnetic field are connected with the current fluctuations (6.53) by means of the relationship

$$\langle \rho^2 \rangle_{\mathbf{k}\omega} = \frac{1}{\omega^2} k_i k_j \langle j_i j_j \rangle_{\mathbf{k}\omega}.\tag{6.59}$$

Using (6.54) we find

$$\langle \rho^2 \rangle_{\mathbf{k}\omega} = \frac{k^2}{2\pi} \frac{T}{\omega} \operatorname{Im} \left\{ 1 - \frac{\Delta(\omega, \mathbf{k})}{\Lambda(\omega, \mathbf{k})} \right\},\tag{6.60}$$

where

$$\Delta(\omega, \mathbf{k}) = \eta^4 - (\epsilon_{11} \cos^2 \vartheta + \epsilon_{22} + \epsilon_{33} \sin^2 \vartheta - 2\epsilon_{13} \sin \vartheta \cos \vartheta)\eta^2$$

$$+ (\epsilon_{11}\epsilon_{22} + \epsilon_{12}^2) \cos^2 \vartheta + (\epsilon_{22}\epsilon_{33} + \epsilon_{23}^2) \sin^2 \vartheta$$

$$+ 2(\epsilon_{12}\epsilon_{23} - \epsilon_{22}\epsilon_{13}) \sin \vartheta \cos \vartheta.\tag{6.61}$$

Let us integrate the spectral distribution (6.60) with respect to the frequencies by using (2.47). Remarking that

$$\frac{\Lambda(0, \mathbf{k})}{\varDelta(0, \mathbf{k})} = 1 + \frac{1}{a^2 k^2},$$

we obtain

$$\langle \rho^2 \rangle_k = \frac{e^2 n_0 k^2}{k^2 + (4\pi \ e^2 n_0/T)}. \tag{6.62}$$

We therefore see that the magnetic field does not affect the spatial correlation function of the plasma density fluctuations when only plasma oscillations are taken into account.

However, the magnetic field exerts a substantial effect on the spectral distribution of the density fluctuations. If wave damping in the plasma is neglected, the spectral distribution of the fluctuations (6.53) becomes

$$\langle j_i j_j \rangle_{\mathbf{k}\omega} = \tfrac{1}{2}\omega T \Lambda^0_{jk} \lambda_{kl} \Lambda^0_{li} \ \delta(A\eta^4 + B\eta^2 + C). \tag{6.63}$$

The argument of the delta-function is the left side of the dispersion equation for a plasma in a magnetic field (6.13). Therefore, only frequencies of the proper oscillations of a plasma in a magnetic field are contained in the fluctuation spectrum in the weak damping case. If $\omega \gg ks$, the spectral distribution of the density fluctuations is determined by the formula (5)

$$\langle \rho^2 \rangle_{\mathbf{k}\omega} = \frac{1}{4} k^2 T \frac{\omega^2}{\Omega^2} \frac{(\omega^2 - \omega_B{}^2)^2}{\omega^4 \sin^2 \vartheta + (\omega^2 - \omega_B{}^2)^2 \cos^2 \vartheta} \{\delta(\omega - \omega_+)$$
$$+ \ \delta(\omega + \omega_+) + \delta(\omega - \omega_-) + \delta(\omega + \omega_-)\}, \tag{6.64}$$

where ω_+ and ω_- are the proper frequencies of the plasma oscillations in a magnetic field (6.29).

The spectral distribution of the electric-field fluctuations in a magnetoactive plasma may be represented, according to (5.81), as

$$\langle E_i E_j \rangle_{\mathbf{k}\omega} = 8\pi^2 e_i{}^* e_j T \left| \frac{\mathrm{Sp}\ \lambda}{\omega} \right| \delta\{\Lambda(\omega, \mathbf{k})\}. \tag{6.65}$$

Remarking that

$$\delta\{\Lambda(\omega, \mathbf{k})\} = \frac{1}{|\ \mathrm{Sp}\ \lambda\ |} \left\{ \frac{1}{|\ \mathbf{e}\ |^2 - \dfrac{|\ \mathbf{ke}\ |^2}{k^2}} \left[\delta(\eta^2 - \eta_+{}^2) + \delta(\eta^2 - \eta_-{}^2)\right] + \delta(\Lambda) \right\},$$

contributions associated with the individual natural oscillations in the plasma may be separated out in (6.65). Thus, for the spectral distributions of fluctuations associated with ordinary and extra-ordinary waves, the formulas

$$\langle E_i E_j \rangle_{\mathbf{k}\omega} = 8\pi^2 \frac{e_i{}^* e_j}{|e|^2 - \frac{|\mathbf{k}e|^2}{k^2}} \frac{T}{\omega} \delta(\eta^2 - \eta_{\pm}{}^2) \tag{6.66}$$

hold, where the polarization vectors are determined by (6.26). The spectral distribution for the Langmuir fluctuations of the electric field is determined by the expression

$$\langle E_i E_j \rangle_{\mathbf{k}\omega} = 4\pi^2 \frac{k_i k_j}{k^2} T \frac{|\omega^2 - \omega_B{}^2|}{\omega_+{}^2 - \omega_-{}^2}$$

$$\times \{\delta(\omega - \omega_+) + \delta(\omega + \omega_+) + \delta(\omega - \omega_-) + \delta(\omega + \omega_-)\} \tag{6.67}$$

where

$$\langle E_i E_j \rangle_{\mathbf{k}\omega} = 8\pi^2 \frac{k_i k_j}{k^2} \frac{T}{\omega} \frac{\eta_L}{A_0} \delta(\eta^2 - \eta_L{}^2), \tag{6.68}$$

if dispersion of the Langmuir waves is taken into account. Formulas (6.65)–(6.68) describe the fluctuations in the high-frequency range of the plasma spectrum, where the influence of the ions is negligible.

REFERENCES

1. V. D. Shafranov, "Electromagnetic Waves in Plasmas." IAE AN SSSR, Moscow, 1960.
2. A. G. Sitenko and K. N. Stepanov, *ZETF* **31**, 642 (1956) *(Sov. Phys.—JETP)*.
3. V. L. Ginzburg, "Electromagnetic Wave Propagation in Plasmas." Fizmatgiz, Moscow, 1960 (Pergamon Press, Oxford).
4. E. P. Gross, *Phys. Rev.* **82**, 232 (1951).
5. A. I. Akhiezer, I. A. Akhiezer, and A. G. Sitenko, *ZETF* **41**, 644 (1961) *(Sov. Phys.—JETP)*.

7
Electron-Ion Plasma in a
Magnetic Field

1. Taking Account of Ion Motion in a Magnetoactive Plasma

In the previous chapter we considered the electromagnetic properties of a plasma in an external magnetic field, where we neglected ion motion completely. However, such an analysis, based on neglecting the ion motion, turns out to be valid only in the high-frequency range; the influence of ions on the electromagnetic properties of a plasma turns out to be very essential in the low-frequency range of the spectrum. In particular, taking account of the ion motion leads to the appearance of characteristic resonance effects in the plasma at frequencies which are multiples of the ion cyclotron frequency, as well as to the possibility of the existence of characteristic low-frequency waves, similar in nature to magnetohydrodynamic waves, in a plasma in a magnetic field.

The influence of ion motion on the electromagnetic properties of plasma is easy to take into account when it is noted that the ion contribution to the dielectric permittivity tensor of the plasma ϵ_{ij} will be expressed by members of the same kind as for electrons, but with an appropriate change in the charge and mass, and in the distribution function, too, in the case of a nonequilibrium plasma. Utilizing (6.18), the dielectric permittivity tensor of an electron–ion plasma in a magnetic field may be written as:

$$\epsilon_{ij}(\omega, \mathbf{k}) = \delta_{ij} - \sum_{\alpha} \frac{\Omega_{\alpha}^{2}}{\omega^{2}}$$

$$\times \left\{ \delta_{ij} - \sum_{n=-\infty}^{\infty} \int d\mathbf{v} \left(\frac{n\omega_B{}^{\alpha}}{v_{\perp}} \frac{\partial f_0{}^{\alpha}}{\partial v_{\perp}} + k_{\parallel} \frac{\partial f_0{}^{\alpha}}{\partial v_{\parallel}} \right) \frac{\Pi_{ij}^{\alpha}(n, \mathbf{v})}{k_{\parallel}v_{\parallel} + n\omega_B{}^{\alpha} - \omega - io} \right\}, \quad (7.1)$$

where

$$\Pi_{ij}^{\alpha}(n, \mathbf{v}) = \begin{pmatrix} \dfrac{n^2\omega_B^{\alpha 2}}{k_{\perp}^{2}} J_n{}^2 & iv_{\perp} \dfrac{n\omega_B{}^{\alpha}}{k_{\perp}} J_n J_n{}' & v_{\parallel} \dfrac{n\omega_B{}^{\alpha}}{k_{\perp}} J_n{}^2 \\ -iv_{\perp} \dfrac{n\omega_B{}^{\alpha}}{k_{\perp}} J_n J_n{}' & v_{\perp}{}^2 J_n'^2 & -iv_{\parallel} v_{\perp} J_n J_n{}' \\ v_{\parallel} \dfrac{n\omega_B{}^{\alpha}}{k_{\perp}} J_n{}^2 & iv_{\parallel} v_{\perp} J_n J_n{}' & v_{\parallel}{}^2 J_n{}^2 \end{pmatrix} \quad (7.2)$$

$$J_n = J_n(a^{\alpha}), \qquad J_n{}' = \frac{\partial}{\partial a^{\alpha}} J_n(a^{\alpha}), \qquad a^{\alpha} = \frac{k_{\perp}v_{\perp}}{\omega_B{}^{\alpha}},$$

and the summation is over the kinds of particles. The tensor (7.1) is written in a coordinate system in which the z axis is along the external magnetic field $\mathbf{B}_0$ and the x axis is in the plane of the vectors $\mathbf{k}$ and $\mathbf{B}_0$.

The dielectric permittivity tensor $\epsilon_{ij}(\omega, \mathbf{k})$ defines the electrodynamic properties of the plasma completely. If an explicit dependence of the dielectric permittivity tensor on the frequency and wave vector is known, then the types of waves being propagated in an unbounded magnetoactive plasma and their characteristics are easily found on the basis of the dispersion equation.

For an equilibrium plasma or a nonisothermal plasma in which the particles are characterized by Maxwellian distributions with different temperatures, the dielectric permittivity tensor is determined by the expression (1, 2)

$$\epsilon_{ij}(\omega, \mathbf{k}) = \delta_{ij} - \sum_{\alpha} \frac{\Omega_{\alpha}^{2}}{\omega^{2}} \left\{ e^{-\beta_{\alpha}} \sum_{n} \frac{z_0{}^{\alpha}}{z_n} \pi_{ij}(z_n{}^{\alpha}) \right.$$

$$\times \left. [\varphi(z_n{}^{\alpha}) - i\sqrt{\pi} z_n{}^{\alpha} \exp(-z_n^{\alpha 2})] - 2z_0^{\alpha 2} b_i b_j \right\}, \quad (7.3)$$

where $\mathbf{b}$ is the unit vector in the $\mathbf{B}_0$ direction,

$$\pi_{ij}(z_n^{\alpha}) = \begin{pmatrix} \dfrac{n^2}{\beta_{\alpha}} I_n & in(I_n' - I_n) \\[2ex] -in(I_n' - I_n) & \left(\dfrac{n^2}{\beta_{\alpha}} + 2\beta_{\alpha}\right) I_n - 2\beta_{\alpha} I_n' \\[2ex] \dfrac{k_{\parallel}}{|k_{\parallel}|} \sqrt{\dfrac{2}{\beta_{\alpha}}} \, n z_n^{\alpha} I_n & i\dfrac{k_{\parallel}}{|k_{\parallel}|} \sqrt{2\beta_{\alpha}}\, z_n^{\alpha}(I_n' - I_n) \end{pmatrix}$$

$$\begin{pmatrix} \dfrac{k_{\parallel}}{|k_{\parallel}|} \sqrt{\dfrac{2}{\beta_{\alpha}}}\, n z_n^{\alpha} I_n \\[2ex] -i\dfrac{k_{\parallel}}{|k_{\parallel}|} \sqrt{2\beta_{\alpha}}\, z_n^{\alpha}(I_n' - I_n) \\[2ex] 2 z_n^{2} I_n \end{pmatrix} ; \qquad (7.4)$$

$I_n = I_n(\beta_{\alpha})$ is the modified Bessel function, $I_n' = \partial/\partial\beta_{\alpha} \, I_n(\beta_{\alpha})$.

$$\Omega_{\alpha}^{2} = \frac{4\pi e^2 h_0}{m_{\alpha}}, \qquad \omega_B^{\alpha} = \frac{e_{\alpha} B_0}{m_{\alpha} c}, \qquad s_{\alpha}^{2} = \frac{3 T_{\alpha}}{m_{\alpha}}, \qquad \sqrt{\beta_{\alpha}} = \frac{k_{\perp} s_{\alpha}}{\sqrt{3}\,\omega_B{}^{\alpha}}$$

$$z_n^{\alpha} = \sqrt{\frac{3}{2}} \, \frac{\omega - n\omega_B^{\alpha}}{|k_{\parallel}|\, s_{\alpha}}, \qquad \alpha = e \text{ or } i, \quad e_e = -e_i = e.$$

As before, the dispersion equation for electromagnetic waves in a plasma taking account of ion motion is determined by (6.20) in which (7.1) or (7.3) should, however, be used.

The spatial dispersion is insignificant when the thermal motion of the electrons and ions is neglected (large wavelength domain) $ks_{\alpha} \ll \omega$, and from (7.3) we obtain well-known expressions for the dielectric permittivity tensor of an electron–ion plasma ϵ_{ij}*:

$$\epsilon_{ij}(\omega) = \begin{pmatrix} \epsilon_1 & -i\epsilon_2 & 0 \\ i\epsilon_2 & \epsilon_1 & 0 \\ 0 & 0 & \epsilon_3 \end{pmatrix}, \qquad (7.5)$$

where

$$\epsilon_1 = 1 - \frac{\Omega_e^{2}}{\omega^2 - \omega_B^{e2}} - \frac{\Omega_i^{2}}{\omega^2 - \omega_B^{i2}},$$

$$\epsilon_2 = \frac{\omega_B^{e}}{\omega} \frac{\Omega_e^{2}}{\omega^2 - \omega_B^{e2}} + \frac{\omega_B^{i}}{\omega} \frac{\Omega_i^{2}}{\omega^2 - \omega_B^{i2}}, \qquad \epsilon_3 = 1 - \frac{\Omega_e^{2}}{\omega^2} - \frac{\Omega_i^{2}}{\omega^2}. \qquad (7.6)$$

In this approximation the tensor (7.5) agrees externally with the

* See, e.g., the Ginzburg book (3).

dielectric permittivity tensor for an electron plasma, determined by (6.22). Hence, the solutions of the dispersion equation for electromagnetic waves in the case of an electron–ion plasma will be determined, as before, by (6.25) and (6.28), in which it is however necessary to take account of the frequency dependence of the ϵ_{ij} components given by (7.6).

In practice, the ϵ_3 component for an electron–ion plasma is not different from the corresponding quantity for the electron plasma, because of the smallness of the ratio between the electron and ion masses m/M. However, the components ϵ_1 and ϵ_2 depend in an essential way on the ion motion. Thus, these components become infinite (when neglecting the thermal motion of the particles) not only at the frequency $\omega = \omega_B{}^e$, but also at $\omega = \omega_B{}^i$. The strong dependence of the electromagnetic properties of a plasma on the ion motion at frequencies on the order of the ion cyclotron frequency is indeed explained by this circumstance. Using (7.6) it is easy to see that taking account of ion motion in a plasma turns out to be essential to the frequency $\omega \sim \sqrt{m/M}\, \omega_B{}^e$. Indeed, in the frequency range $|\omega_B{}^i| \ll \omega < \sqrt{m/M}\, \omega_B{}^e$ the expressions (7.6) are

$$\epsilon_1 = 1 + \frac{\Omega_e{}^2}{\omega_B^{e2}} - \frac{\Omega_i{}^2}{\omega^2}, \qquad \epsilon_2 = \frac{\Omega_e{}^2}{\omega \omega_B{}^e}, \qquad \epsilon_3 = 1 - \frac{\Omega_e{}^2}{\omega^2}.$$

It hence follows directly that the ion motion will affect electromagnetic wave propagation in the plasma up to frequencies $\sqrt{m/M}\, \omega_B{}^e$. If $\omega \gg \sqrt{m/M}\, \omega_B{}^e$, the member $\Omega_i{}^2/\omega^2$ in the expression for ϵ_1 may be neglected; hence the electromagnetic properties of a plasma turn out to be independent of the ions.

2. Ion Cyclotron Resonance

Let us consider the electromagnetic properties of an electron–ion plasma in a frequency band close to the ion cyclotron frequency $\omega_B{}^i = -(eB_0/Mc)$ [Shafranov (4)]. Retaining only the main terms in (7.6) and considering the particle density in the plasma to be sufficiently great $\Omega_e \gg \omega$, we have

$$\epsilon_1 = -\frac{\Omega_i{}^2}{\omega^2 - \omega_B^{i2}}, \qquad \epsilon_2 = \frac{\omega_B{}^i}{\omega}\frac{\Omega_i{}^2}{\omega^2 - \omega_B^{i2}}, \qquad \epsilon_3 = -\frac{M}{m}\frac{\Omega_i{}^2}{\omega^2}. \qquad (7.7)$$

Substituting (7.7) into (6.25) and taking account of the smallness of the ratio m/M, we obtain the following expressions for the refractive indices of electromagnetic waves in a plasma in the domain of ion cyclotron resonance:

$$\eta_{\pm}^2 = \frac{\epsilon_1(1 + \cos^2 \vartheta) \pm \sqrt{\epsilon_1^2 \sin^4 \vartheta + 4\epsilon_2^2 \cos^2 \vartheta}}{2 \cos^2 \vartheta} \qquad (7.8)$$

For $\omega \to |\omega_B{}^i|$ the refractive index of the ordinary wave η_{+}^2 becomes infinite. This means that intensive electromagnetic wave absorption by the plasma ions occurs in the mentioned frequency range. (The refractive index of the extraordinary wave η_{-}^2 remains finite for $\omega \to |\omega_B{}^i|$.) Formula (7.8) has been obtained by neglecting the thermal motion of the plasma particles. To find the coefficient of electromagnetic-wave absorption by the plasma ions it is necessary to take into account thermal corrections to ϵ_{ij}, which will be particularly essential near cyclotron resonance. Utilizing (7.3), we easily obtain the following expression for the imaginary additions to the quantities ϵ_1 and ϵ_2 near ion cyclotron resonance:

$$\epsilon_1'' = \epsilon_2'' = \sqrt{\frac{\pi}{8}} \frac{\Omega_i}{a_i k \omega |\cos \vartheta|} \exp\left[-\frac{3}{2} \frac{(|\omega_B{}^i| - \omega)^2}{k^2 s_i^2 \cos^2 \vartheta}\right]. \qquad (7.9)$$

[Inessential electron members have been omitted from (7.9).] Neglecting to take account of the thermal motion in the real parts of ϵ_1 and ϵ_2, and linearizing (7.6) with respect to the imaginary additions to ϵ_1 and ϵ_2, we obtain the following expressions for the absorption coefficients (imaginary parts of η) of the electromagnetic waves in a magnetoactive plasma in the domain of ion cyclotron resonance:

$$\eta_{\pm}'' = \frac{1 + \cos^2 \vartheta \pm (\epsilon_1' \sin^4 \vartheta + 4\epsilon_2' \cos^2 \vartheta)(\epsilon_1'^2 \sin^4 \vartheta + 4\epsilon_2'^2 \cos^2 \vartheta)^{1/2}}{2\eta_{\pm}' \cos^2 \vartheta} \epsilon_1'',$$

$$(7.10)$$

where $\eta_{\pm}'$ are the refractive indices of the waves determined by the relations (7.8). It follows from (7.10) that absorption occurs in the cyclotron resonance domain for any direction of wave propagation.

3. Magnetohydrodynamic Waves

As is known, magnetohydrodynamic waves (Alfven wave and two magnetosonic waves) may be propagated in an ideally conducting fluid in a magnetic field. These waves are characterized by the following dispersion laws:

$$\omega^2 = k^2 v_A^2 \cos^2 \vartheta, \tag{7.11}$$

$$\omega^2 = \tfrac{1}{2}k^2\{v_A^2 + v_S^2 \pm [(v_A^2 + v_S^2)^2 - 4v_A^2 v_S^2 \cos^2 \vartheta]^{1/2}\}, \tag{7.12}$$

where ϑ is the angle between the wave propagation direction $\mathbf{k}$ and the magnetic field $\mathbf{B_0}$, $v_A = B_0/(4\pi n_0 M)^{1/2}$ is the Alfven velocity, and v_S the speed of sound in the fluid. The relations (7.11) and (7.12) are valid for $v_A \ll c$ (c the velocity of light).

It turns out that waves of similar nature may be propagated also in an electron–ion plasma in a magnetic field, at low frequencies compared with the ion cyclotron frequency $\omega_B{}^i$. These low-frequency waves in a plasma are also customarily designated magnetohydrodynamic waves [Ginzburg (5)].*

Let us consider low-frequency waves in a plasma by considering the magnetic field sufficiently strong $ks_\alpha \ll \omega_B{}^\alpha$. Expanding the function $I_n(\beta_\alpha)$ and $e^{-\beta_\alpha}$ in (7.3) in power series in β_α, we obtain the following approximate expressions for the components of the ϵ_{ij} tensor in the low-frequency range $\omega \ll \omega_B{}^\alpha$:

$$\epsilon_{11} = \epsilon_0, \qquad \epsilon_{12} = i\epsilon_0 \frac{\omega}{\omega_B{}^i}\left[1 - \frac{1}{2}\frac{k^2 s_i^2}{\omega^2}\left(\sin^2\vartheta - \frac{2}{3}\cos^2\vartheta\right)\right], \qquad \epsilon_{13} = 0$$

$$\epsilon_{22} = \epsilon_0 \left\{1 - \frac{2}{3}\frac{k^2 s_i^2}{\omega^2}[t\varphi(z) + \varphi(\mu z) - i\sqrt{\pi}z(te^{-z^2} + \mu e^{-\mu^2 z^2})]\sin^2\vartheta\right\},$$

$$\tag{7.13}$$

$$\epsilon_{23} = i\epsilon_0 \frac{\omega_B{}^i}{\omega}[\varphi(z) - \varphi(\mu z) - i\sqrt{\pi}z(e^{-z^2} - \mu e^{-\mu^2 z^2})]\tan\vartheta,$$

$$\epsilon_{33} = 3\epsilon_0 \frac{\omega_B{}^{i2}}{k^2 s_i^2}\left[\frac{1-\varphi(z)}{t} + 1 - \varphi(\mu z) + i\sqrt{\pi}z\left(\frac{1}{t}e^{-z^2} + \mu e^{-\mu^2 z^2}\right)\right]\frac{1}{\cos^2\vartheta},$$

---

* Braginskii and Kazantsev (6) examined magnetohydrodynamic waves in plasma. The question of the applicability of a magnetohydrodynamic description for a rarefied plasma without collisions is discussed by Stepanov (7), Klimontovich and Silin (8).

where

$$\epsilon_0 = \frac{\Omega_i{}^2}{\omega_B{}^{i2}}, \qquad t = \frac{T_e}{T_i}, \qquad \mu = t\frac{M}{m}, \qquad \text{and} \qquad z = \sqrt{\frac{3}{2}}\frac{\omega}{ks}.$$

(Let us also present the expressions for the components of the electron part of the plasma susceptibility χ_{ij}^e in the low-frequency range $\omega \ll \omega_B{}^\alpha$:

$$\chi_{11}^e = \frac{\epsilon_0}{4\pi}\frac{m}{M},$$

$$\chi_{12}^e = -i\,\frac{\epsilon_0}{4\pi}\frac{\omega_B{}^i}{\omega}, \qquad \chi_{13}^e \simeq 0,$$

$$\chi_{22}^e = \frac{\epsilon_0}{4\pi}\frac{m}{M}\left\{1 - \frac{2}{3}\frac{k^2s^2}{\omega^2}\left[\varphi(z) - i\sqrt{\pi}ze^{-z^2}\right]\sin^2\vartheta\right\}, \qquad (7.14)$$

$$\chi_{23}^e = i\,\frac{\epsilon_0}{4\pi}\frac{\omega_B{}^i}{\omega}\left[\varphi(z) - 1 - i\sqrt{\pi}ze^{-z^2}\right]\tan\vartheta,$$

$$\chi_{33}^e = 3\,\frac{\epsilon_0}{4\pi}\frac{\omega_B{}^{i2}}{k^2s_i{}^2}\frac{1}{t}\left[1 - \varphi(z) + i\sqrt{\pi}z\,e^{-z^2}\right]\frac{1}{\cos^2\vartheta}\cdot\bigg)$$

Substituting (7.13) into (6.20), the dispersion equation may be represented as

$$(\eta^2\cos^2\vartheta - \epsilon_{11})[(\eta^2 - \epsilon_{22})\,\epsilon_{33} - \epsilon_{23}^2] + 2\eta^2\epsilon_{12}\epsilon_{23}\sin\vartheta\cos\vartheta$$

$$+ \epsilon_{12}^2\epsilon_{33} + \cdots = 0. \qquad (7.15)$$

The first member in (7.15) contains terms proportional to $(\omega_B{}^i/\omega)^2$ or $(\omega_B{}^i/ks_i)^2$. The second and third members in (7.15) are independent of the parameters $(\omega/\omega_B{}^i)^2$ and $(ks_i/\omega_B{}^i)^2$. The subsequent members, proportional to higher powers of the small parameters $(\omega/\omega_B{}^i)^2$ and $(ks_i/\omega_B{}^i)^2$, have been omitted in (7.15).

If the free terms in (7.15) are neglected, it will decompose into two independent equations:

$$\eta^2\cos^2\vartheta - \epsilon_{11} = 0, \qquad (\eta^2 - \epsilon_{22})\,\epsilon_{33} - \epsilon_{23}^2 = 0. \qquad (7.16)$$

The first equation in (7.16) determines the refractive index of the ordinary wave

$$\eta^2 = \frac{\Omega_i{}^2}{\omega_B^{i2}} \frac{1}{\cos^2 \vartheta}. \tag{7.17}$$

Expressing η in terms of k and ω, the relationship (7.17) may be rewritten as

$$\omega^2 = k^2 v_A{}^2 \cos^2 \vartheta, \qquad v_A = \frac{B_0}{(4\pi n_0 M)^{1/2}}. \tag{7.18}$$

Therefore, an ordinary wave in an electron–ion plasma in a magnetic field is a magnetohydrodynamic Alfven wave in the low-frequency range. Taking account of the free terms in (7.15) we may find the thermal corrections to the phase velocity and damping of the Alfven waves. If $(s_i{}^2/c^2)(\Omega_i{}^2/\omega_B^{i2}) \ll 1$, then the damping is determined by the expression

$$\frac{\gamma}{\omega} = \frac{1}{2}\sqrt{\frac{\pi}{6}} \frac{m}{M} \frac{s}{v_A} \frac{\omega^2}{\omega_B^{i2}} (\tan^2 \vartheta + \cotan^2 \vartheta), \qquad s_i \ll v_A \ll s. \tag{7.19}$$

The polarization vector for Alfven waves equals

$$\mathbf{e}_A = \left(1; \quad -i \frac{\omega}{\omega_B{}^i} \cotan^2 \vartheta; \quad -\frac{v_s{}^2}{v_A{}^2} \frac{\omega^2}{\omega_B^{i2}} \frac{1}{\sin \vartheta \cos \vartheta}\right). \tag{7.20}$$

The second equation of (7.16) determines the refractive index of the extraordinary wave. The roots of (7.16) depend in an essential manner on the relationships between the thermal velocities of the electrons and ions s and s_i and the Alfven velocity v_A. If $s_i{}^2 \ll v_A{}^2 \ll s^2$, then by using (7.13) we find $\eta^2 = \Omega_i{}^2/\omega_B^{i2}$ in a first approximation, from (7.16), or

$$\omega^2 = k^2 v_A{}^2. \tag{7.21}$$

Thermal corrections to (7.21) may be found by successive approximations. Since $\mu z \gg 1$ for all angles ϑ, and $z \ll 1$ in the range of angles for which $\cos \vartheta \gg v_A/s$, we then find in a second approximation

$$\omega^2 = k^2 v_A{}^2 + \tfrac{2}{3}(1 + \tfrac{1}{2}t)k^2 s_i{}^2 \sin^2 \vartheta. \tag{7.22}$$

The wave damping equals

$$\frac{\gamma}{\omega} = \frac{1}{2}\sqrt{\frac{\pi}{6}} \frac{m}{M} \frac{s}{v_A} \frac{\sin^2 \vartheta}{|\cos \vartheta|}. \tag{7.23}$$

The expression (7.22) is valid for all angles except those very close to $\pi/2$. For angles close to $\pi/2$ ($\cos \vartheta \ll v_A/s$), we find approximately from (7.16):

$$\omega^2 = k^2 v_A{}^2 + \tfrac{2}{3}(1 + t)\, k^2 s_i{}^2. \tag{7.24}$$

Comparing (7.22) or (7.24) with the corresponding root of (7.12), which equals $\omega^2 = k^2 v_A{}^2 + k^2 v_s{}^2 \sin^2 \vartheta$ for $v_s \ll v_A$, we see that the extraordinary wave in an electron–ion plasma in the presence of a magnetic field may be considered as a magnetosonic wave in the low-frequency range if the velocity of sound is defined by means of the equalities

$$v_s{}^2 = \tfrac{2}{3}(1 + \tfrac{1}{2}t)\, s_i{}^2, \qquad \cos \vartheta \gg \frac{v_A}{s},$$

$$v_s{}^2 = \tfrac{2}{3}(1 + t)\, s_i{}^2, \qquad \cos \vartheta \ll \frac{v_A}{s}.$$

The polarization vector of a fast magnetosonic wave is determined by the expression

$$\mathbf{e}_M = \left(-i\, \frac{\omega}{\omega_B{}^i}\, \frac{1}{\sin^2 \vartheta}\; ;\quad 1;\quad -i\, \frac{v_s{}^2}{v_A{}^2}\, \frac{\omega}{\omega_B{}^i} \sin \vartheta \cos \vartheta\right). \tag{7.25}$$

Let us note that nonisothermal sound waves similar to (5.14) may also be propagated in a strongly nonisothermal plasma $T_e \gg T_i$, in a magnetic field. Assuming that $s_i \ll \omega/(k \cos \vartheta) \ll s$, the following solution may be obtained from (7.16):

$$\eta^2 = \frac{c^2}{v_s{}^2}\, \frac{1}{\cos^2 \vartheta}, \qquad \omega^2 = k^2 v_s{}^2 \cos^2 \vartheta, \qquad v_s{}^2 = \frac{T_e}{M}. \tag{7.26}$$

Comparing (7.26) with (5.14), we see that the magnetic field leads to the dependence of the velocity of nonisothermal sound on the direction of wave propagation.

The damping of nonisothermal sound is

$$\frac{\gamma}{\omega} = \left(\frac{\pi}{8}\, \frac{m}{M}\right)^{1/2}. \tag{7.27}$$

The polarization vector is determined by the expression

$$\mathbf{e}_s = \left(\sin \vartheta;\quad -i\, \frac{v_s{}^2}{v_A{}^2}\, \frac{\omega_B{}^i}{\omega} \sin \vartheta \cos^2 \vartheta;\quad \cos \vartheta\right). \tag{7.28}$$

4. Low-Frequency Fluctuations in a Magnetoactive Plasma

The possibility of the existence of weakly damped magneto-hydrodynamic oscillations in a plasma in an external magnetic field results in the appearance of additional maximums, associated with these oscillations, in the low-frequency portion of the fluctuation spectrum. The fluctuations associated with Alfven and magnetosonic oscillations in a magnetoactive plasma are described by the general formulas (5.57)–(5.59) and (5.63), in which (7.13) and (7.14) should be used for the dielectric susceptibilities. Retaining the highest powers of the large parameter $\omega_B{}^i/\omega$, simple formulas may be obtained for the field correlators for $\omega \ll \omega_B{}^i$.

Thus, the spectral distribution of the electric field fluctuations near the Alfven frequencies $\omega = \pm k v_A \cos \vartheta$ may be represented in the form:

$$\langle E_i E_j \rangle_{\mathbf{k}\omega} = 4\pi^2 e_i{}^* e_j T_e \frac{v_A{}^2}{c^2} \{\delta(\omega - k v_A \cos \vartheta) + \delta(\omega + k v_A \cos \vartheta)\},$$

$$(7.29)$$

where the vector $\mathbf{e}_A$ is defined by (7.20). The magnetosonic fluctuations of the electric field are characterized by the spectral distributions:

$$\langle E_i E_j \rangle_{\mathbf{k}\omega} = 4\pi^2 e_i{}^* e_j T_e \frac{v_A{}^2}{c^2} \{\delta(\omega - k v_A) + \delta(\omega + k v_A)\}, \qquad (7.30)$$

$$\langle E_i E_j \rangle_{\mathbf{k}\omega} = 4\pi^2 e_i{}^* e_j T_e a^2 k^2 \{\delta(\omega - k v_s \cos \vartheta) + \delta(\omega + k v_s \cos \vartheta)\}, \quad (7.31)$$

where the vectors $\mathbf{e}_M$ and $\mathbf{e}_s$ equal (7.25) and (7.28), respectively. Let us note that the spectral distributions for the low-frequency fluctuations are proportional to the square of the ratio between the phase velocity of the corresponding wave and the velocity of light in vacuo.

5. Fluctuations in a Nonequilibrium Magnetoactive Plasma

Let us consider a plasma through which a compensated beam of charged particles passes with velocity $\mathbf{u}$ directed along the magnetic field. If the particle distribution in the plasma at rest and in the beam are Maxwellian with temperatures T and T' the components of the dielectric permittivity tensor will equal

$$\epsilon_{ij}(\omega, \mathbf{k}) = \delta_{ij} - \sum \frac{\Omega^2}{\omega^2}\left\{ e^{-\beta} \sum_n \frac{z_0}{z_n} \pi_{ij}(z_n)[\varphi(z_n) - i\sqrt{\pi}\, z_n e^{-z_n^2}] - 2z_0^2 b_i b_j \right\}$$

$$- \sum \frac{\Omega'^2}{\omega^2}\left\{ e^{-\beta'} \sum_n \frac{y_0}{y_n} \pi_{ij}(z_n')[\varphi(y_n) - i\sqrt{\pi}\, y_n e^{-y_n^2}] - 2z_0'^2 b_i b_j \right\}, \qquad (7.32)$$

where

$$y_n = \sqrt{\frac{3}{2}} \frac{\omega - n\omega_B - k_\parallel u}{|k_\parallel|s'}.$$

The primes denote quantities referring to the beam.

The spectral distribution of the current fluctuations of the free particles is defined by

$$\langle j_i j_j \rangle_{\mathbf{k}\omega}^0 = \sqrt{\frac{2\pi}{3}} \frac{e^2}{|k_\parallel|}\left\{ \sum n_0 s e^{-\beta} \sum_n \pi_{ji}(z_n)e^{-z_n^2} \right.$$

$$\left. - \sum n_0' s' e^{-\beta'} \sum_n \pi_{ji}(z_n')e^{-y_n^2} \right\}. \qquad (7.33)$$

Utilizing (7.32) and (7.33), the spectral distribution of the field fluctuations in a plasma-beam system may be found according to (5.8). We shall henceforth limit ourselves to the consideration of a low density ($n_0' \ll n_0$) beam. In this case the influence of the beam on the wave dispersion in the plasma may be neglected. However, the influence of the beam will be exerted substantially on the effective temperature of the fluctuation oscillations.

The thermal motion of the particles may also be neglected when considering wave dispersion in the high-frequency domain. The spectral distribution of the field fluctuations will hence be determined by the formula

$$\langle E_i E_j \rangle_{\mathbf{k}\omega} = 8\pi^2 \frac{\tilde{T}}{|\omega|}\left\{ \frac{e_i^* e_j}{|\mathbf{e}|^2 - \dfrac{|\mathbf{ke}|^2}{k^2}} [\delta(\eta^2 - \eta_+^2) + \delta(\eta^2 - \eta_-^2)] \right.$$

$$\left. + \frac{k_i k_j}{k^2} \delta(A) \right\}. \qquad (7.34)$$

Since there is no wave damping in a cold plasma, it is necessary to take account of both the thermal motion of the electrons in the plasma, and the presence of the beam when calculating the effective temperature in (7.34). The thermal motion of the ions in both the beam and the plasma at rest may be neglected.

Using (7.32) and (7.33), the effective temperature for high-frequency fluctuation oscillations may be represented as

$$\check{T} = T \frac{R(\omega, \mathbf{k})}{\left| 1 - \dfrac{u}{\tilde{u}} \cos \vartheta \right|}, \tag{7.35}$$

where

$$R(\omega, \mathbf{k}) = \frac{1 + \dfrac{n_0'}{n_0} \left(\dfrac{T}{T'}\right)^{1/2} K}{1 + \dfrac{n_0'}{n_0} \left(\dfrac{T}{T'}\right)^{3/2} K}, \qquad \tilde{u} = \frac{\omega}{k} \left[1 + \frac{n_0}{n_0'} \left(\frac{T'}{T}\right)^{3/2} K^{-1} \right]. \tag{7.36}$$

In the case of fluctuating Langmuir oscillations the quantity K equals

$$K = e^{\beta - \beta'} \frac{\sum_n I_n(\beta') e^{-y_n^2}}{\sum_n I_n(\beta) e^{-z_n^2}}. \tag{7.37}$$

For fluctuating ordinary and extraordinary electromagnetic waves we have

$$K = e^{\beta - \beta'} \frac{\sum_n \left\{ \left[\rho'^2 + (n^2 + \beta'^2) \dfrac{\epsilon_2^2}{(\eta^2 - \epsilon_1)^2} \right] I_n(\beta') - 2\beta'\rho' \dfrac{\epsilon_2^2}{\eta^2 - \epsilon_1} I_n'(\beta') \right\} e^{-y_n^2}}{\sum_n \left\{ \left[\rho^2 + (n^2 + \beta^2) \dfrac{\epsilon_2^2}{(\eta^2 - \epsilon_1)^2} \right] I_n(\beta) - 2\beta\rho \dfrac{\epsilon_2^2}{\eta^2 - \epsilon_1} I_n'(\beta) \right\} e^{-z_n^2}}, \tag{7.38}$$

$$\rho = n + \beta \frac{\epsilon^2}{\eta^2 - \epsilon_1} + \sqrt{2\beta} \frac{k_\parallel}{|k_\parallel|} \frac{\eta^2 \sin \vartheta \cos \vartheta}{\eta^2 \sin^2 \vartheta - \epsilon_3} z_n.$$

The quantity $\tilde{u}$ plays the part of the critical velocity, upon reaching which the fluctuations in the plasma increase without limit and the plasma becomes unstable.

In the limiting case $\eta^2 \gg 1$ the expression (7.38) agrees with (7.37). Let us also note that for $\eta^2 \gg 1$ the spectral distribution for the fluctuating ordinary or extraordinary waves agrees with the distribution for the fluctuating Langmuir oscillations.

The effective temperature for the low-frequency fluctuating oscillations in a magnetoactive plasma permeated by a beam of charged particles is defined by

$$\tilde{T} = \frac{T}{|\, 1 - (u/\tilde{u})\cos\vartheta\,|}, \tag{7.39}$$

in which the critical velocities for the Alfven and fast and slow magnetosonic fluctuations equal, respectively,

$$\tilde{u}_A = \frac{\omega}{k} \left\{ 1 + \frac{n_0}{n_0'} \left(\frac{T}{T'}\right)^{1/2} \frac{\sin^4\vartheta + \cos^4\vartheta}{\cos^4\vartheta + [(T/T') - \cos^2\vartheta]^2} e^{y_0^2} \right\}, \tag{7.40}$$

$$\tilde{u}_M = \frac{\omega}{k} \left\{ 1 + \frac{n_0}{n_0'} \sqrt{\frac{T}{T'}} \frac{e^{y_0^2}}{1 + [1 - (T/T')]^2} \right\},$$

$$\tilde{u}_s = \frac{\omega}{k} \left\{ 1 + \frac{n_0}{n_0'} \left(\frac{T'}{T}\right)^{3/2} e^{y_0^2} \right\}. \tag{7.41}$$

The spectral distributions of the low-frequency fluctuations in a nonequilibrium plasma will be determined by (7.29)–(7.31), in which the effective temperature (7.39) should be substituted in place of T_e.

REFERENCES

1. A. G. Sitenko and K. N. Stepanov, *ZETF* **31**, 642 (1956) *(Sov. Phys. JETP)*.
2. K. N. Stepanov, *ZETF* **35**, 1155 (1958) *(Sov. Phys. JETP)*.
3. V. L. Ginzburg, "Electromagnetic Wave Propagation in Plasmas." Fizmatgiz, Moscow, 1960 (Gordon & Breach, New York or Pergamon Press, Oxford).
4. V. D. Shafranov, "Plasma Physics and the Problem of Controlled Thermonuclear Reactions," Vol. 4. AN SSSR Press, Moscow, 1958 (Pergamon Press).
5. V. L. Ginzburg, *ZETF* **21**, 788 (1951).
6. S. I. Braginskii and A. P. Kazantsev, "Plasma Physics and the Problem of Controlled Thermonuclear Reactions," Vol. 4. AN SSSR Press, Moscow, 1958 (Pergamon Press).
7. K. N. Stepanov, *Ukr. Fiz. Zh.* **4**, 678 (1959).
8. Iu. L. Klimontovich and V. P. Silin, *ZETF* **40**, 1213 (1961) *(Sov. Phys. JETP)*.
9. A. G. Sitenko and Iu. A. Kirochkin, *Usp. Fiz. Nauk* **89**, 227 (1966) *(Sov. Phys.— Uspekhi)*.

8

Passage of Charged
Particles through Plasma

1. Field of Charge in a Plasma. Shielding

A peculiarity of the electromagnetic properties of a plasma is also manifest in the analysis of the field and dynamics of a charged particle moving in the plasma. Polarization, manifested in the spatial shielding of the field of charged particles, occurs because of the interaction between the charged particle and the surrounding particles. Moreover, the interaction leads to the origin of a retarding force acting on the moving particle and associated with the excitation of electromagnetic waves in the plasma.

Let us first consider the field generated by the charged particle moving in the plasma. This field is determined by the system of Maxwell equations (2.1) in which the additional charge and current densities associated with the moving particle must be taken into account:

$$\rho_0 = q\delta(\mathbf{r} - \mathbf{v}t), \qquad \mathbf{j}_0 = q\mathbf{v}\delta(\mathbf{r} - \mathbf{v}t), \qquad (8.1)$$

where q is the charge and $\mathbf{v}$ is the velocity of the particle.

Expanding the field and currents into Fourier integrals and eliminating the magnetic induction from the equations, we obtain the following equation for the electric field intensity:

$$\left\{ \eta^2 \left(\frac{k_i k_j}{k^2} - \delta_{ij} \right) + \epsilon_{ij}(\omega, \mathbf{k}) \right\} E_j(\mathbf{k}, \omega) = \frac{4\pi}{i\omega} j_i^0(\mathbf{k}, \omega), \qquad (8.2)$$

where $j_i{}^0(\mathbf{k}, \omega)$ is the Fourier component of the current density, connected with the moving charge. In the case of an isotropic plasma for which the dielectric permittivity tensor is determined by (2.7), the solution of (8.2) is

$$E_i(\mathbf{k}, \omega) = \frac{4\pi}{i\omega} \left\{ \frac{1}{\epsilon_l(\omega, \mathbf{k})} \frac{k_i k_j}{k^2} + \frac{1}{\epsilon_t(\omega, \mathbf{k}) - \eta^2} \left(\delta_{ij} - \frac{k_i k_j}{k^2}\right)\right\} j_j{}^0(\mathbf{k}, \omega).$$

(8.3)

If the velocity of charge motion is constant, the Fourier component of the current density (8.1) equals

$$\mathbf{j}_0(\mathbf{k}, \omega) = \frac{q}{(2\pi)^3} \mathbf{v}\delta(\omega - \mathbf{k}\mathbf{v}).$$

(8.4)

Substituting (8.4) into (8.3), and using the inverse Fourier transform, we obtain the following expression for the electric field of a point charge moving with constant velocity in the plasma:

$$\mathbf{E}(\mathbf{r}, t) = \frac{q}{2\pi^2 i} \int d\mathbf{k} \left\{ \frac{\mathbf{k}}{k^2 \epsilon_l(\omega, \mathbf{k})} + \frac{\mathbf{v} - \dfrac{\mathbf{k}(\mathbf{k}\mathbf{v})}{k^2}}{\omega\left[\epsilon_t(\omega, \mathbf{k}) - \dfrac{k^2 c^2}{\omega^2}\right]}\right\} e^{i\mathbf{k}\mathbf{r} - i\omega t},$$

(8.5)

$$\omega = \mathbf{k}\mathbf{v}.$$

The first member in (8.5) describes the potential (longitudinal) part of the field of the moving charge, the second member in (8.5) corresponds to the vortex (transverse) portion of the field. The vortical portion of the field is directly connected with the charge motion.

If the charge is at rest ($v = 0$), the vortical part in (8.5) is missing, and the field is completely potential. Introducing the scalar field potential φ in this case

$$\mathbf{E} = -\operatorname{grad} \varphi,$$

we find according to (8.5)

$$\varphi(r) = \frac{q}{2\pi^2} \int d\mathbf{k} \, \frac{e^{i\mathbf{k}\mathbf{r}}}{k^2 \epsilon_l(0, \mathbf{k})}.$$

(8.6)

Substituting the static value of the longitudinal dielectric permittivity of the plasma $\epsilon_l(0, \mathbf{k}) = 1 + (1/a^2k^2)$ into (8.6) and integrating, we finally obtain the following expression for the potential of the field of a charge at rest in a plasma

$$\varphi(r) = \frac{q}{r} e^{-r/a}. \tag{8.7}$$

(The charge is assumed to be at the origin.)

As we see, the presence of spatial dispersion in the plasma leads to shielding of the point-charge field. The Debye radius a determines the range at which essential attenuation of the field of a charge at rest in a plasma will occur.

2. Energy Losses in Charged-Particle Motion in a Plasma

Upon passing through a plasma, a charged particle loses part of its energy because of the interaction with surrounding particles, resulting in polarization of the plasma.* These energy losses are simply computed if it is assumed that the energy lost by the particle per unit time is small as compared with the energy of the particle itself, and therefore, the change in velocity of the particle during motion may be neglected.

The energy losses of a charged particle are determined by the work of the retarding forces acting on the particle in the plasma from the electromagnetic field generated by the particle itself while moving. Evidently the energy losses of the particle per unit time are

$$\frac{\partial E}{\partial t} = q\mathbf{v}\mathbf{E} \mid_{\mathbf{r}=vt}, \tag{8.8}$$

where the field is taken at the location of the particle. Using (8.5) for the field, we find

$$\frac{\partial E}{\partial t} = \frac{q^2}{2\pi^2 i} \int d\mathbf{k} \left\{ \frac{\omega}{k^2\epsilon_l(\omega, \mathbf{k})} + \frac{v^2 - (\omega^2/k^2)}{\omega[\epsilon_t(\omega, \mathbf{k}) - (k^2c^2/\omega^2)]} \right\}, \quad \omega = \mathbf{k}\mathbf{v}. \tag{8.9}$$

Integration with respect to $\mathbf{k}$ in (8.9) should be up to some maximum value $\mathbf{k}_0$, for which the macroscopic analysis is still valid.

---

* Passage of a charged particle through a plasma is considered in Reference 1.

Taking into account that the real parts of ϵ_l and ϵ_t are even functions of the frequency, and the imaginary parts are odd functions, (8.9) may be rewritten as:

$$\frac{\partial E}{\partial t} = \frac{q^2}{2\pi^2} \operatorname{Im} \int d\mathbf{k} \left\{ \frac{\omega}{k^2 \epsilon_l(\omega, \mathbf{k})} + \frac{v^2 - (\omega^2/k^2)}{\omega[\epsilon_t(\omega, \mathbf{k}) - (k^2 c^2/\omega^2)]} \right\}, \quad \omega = \mathbf{kv}.$$
(8.10)

The first member in (8.10) takes account of the particle inter-action with the longitudinal field and determines the polarization energy losses of the particle. The second member in (8.10) takes account of particle interaction with the transverse field, and corre-sponds in the general case to energy losses connected with radiation of transverse waves (Cerenkov radiation). Since the phase velocities of the transverse waves are greater than the velocity of light for the isotropic plasma, the Cerenkov-radiation condition is not satisfied, and the second member in (8.10) vanishes in this case.

Let us note that both the polarization energy losses (if recoil of the particle is neglected) as well as the losses associated with the Cerenkov radiation are independent of the mass of the moving particle. In particular, these losses will differ from zero even for particles with infinite mass.

3. Polarization Energy Losses of a Fast Particle in a Plasma

Let us evaluate the polarization energy losses of a charged particle moving with velocity v, significantly greater than the mean thermal velocity of the plasma electrons s. Noting that $d\mathbf{k} = 2\pi k \, dk \, d\omega/v$, where k is the component of $\mathbf{k}$ perpendicular to $\mathbf{v}$, let us rewrite the expression for the polarization losses in the form:

$$\frac{\partial E}{\partial t} = \frac{q^2}{\pi v} \operatorname{Im} \int_0^{k_0} dk \, k \int_{-\infty}^{\infty} d\omega \, \omega \bigg/ \left\{ k^2 + \frac{\omega^2}{v^2} + \frac{1}{a^2} [1 - \varphi(z) + i \sqrt{\pi} z \, e^{-z^2}] \right\}.$$
(8.11)

The expression (8.11) diverges as $k \to \infty$. This is associated with the fact that by describing the plasma by the macroscopic dielectric permittivities we actually take into account only the far interaction, and do not take account of the nearby collisions of the particle with the plasma electrons. These collisions cannot be taken into account if the collision parameter b is considerably greater than v/Ω; if $b \ll v/\Omega$, then collisions of the charged particle with the individual electrons of the plasma play the principal part, and the macroscopic description of the interaction between the particle and the plasma becomes meaningless. Large k correspond to small b, hence the integration with respect to k in (8.11) should be carried out to a maximum value k_0.

If the particle velocity v is considerably greater than the mean thermal velocity of the plasma electrons s, then values $z \gg 1$ are effective. Using the asymptotic expansion (3.28), the energy losses due to long-range collisions may be represented as

$$\frac{\partial E}{\partial t} = \frac{q^2 \Omega^2}{\pi v} \operatorname{Im} \int_0^{k_0} dk\, k \int_{-\infty}^{\infty} \frac{d\omega\, \omega}{[k^2 + (\omega^2/v^2)](\omega^2 - \Omega^2 + 2i\omega\gamma)}, \qquad (8.12)$$

where γ is the damping decrement of the plasma waves (3.34). The integral with respect to ω along the real axis is equal to the residue relative to the single pole $\omega = ikv$, in the upper half-plane, hence

$$\frac{\partial E}{\partial t} = -\frac{q^2 \Omega^2}{v} \int_0^{k_0} \frac{dk\, k}{k^2 + (\Omega^2/v^2)}.$$

Assuming $k_0^2 v^2 \gg \Omega^2$, we obtain the following expression for the energy loss due to the long-range collisions:

$$\frac{\partial E}{\partial t} = -\frac{4\pi n_0 e^2 q^2}{mv} \ln \frac{v k_0}{\Omega}. \qquad (8.13)$$

Let b_0 denote the critical value of the collision parameter separating the domains of short and long-range collisions. It is easy to see that the maximum value k_0 is connected with the critical collision parameter b_0. Indeed, the energy losses due to the long-range collisions may be defined as the energy flux of the electromagnetic field passing through a cylindrical surface of radius b_0 surrounding the trajectory of the charge:

$$\frac{\partial E}{\partial t} = \frac{c}{4\pi} \int [\mathbf{E}, \mathbf{B}]\, ds. \qquad (8.14)$$

Taking account of axial symmetry of the field we may rewrite this equality as:

$$\frac{\partial E}{\partial t} = -\frac{1}{2} b_0 c \int_{-\infty}^{\infty} E_z B \, dz. \tag{8.15}$$

For $v \gg s$ the spatial dispersion is insignificant. In this case Cerenkov radiation is absent, hence, the energy losses (8.15) will be associated only with the longitudinal portion of the electric field. Using (8.5), the component of the longitudinal electric field and the magnetic field may be represented as

$$E_z{}^l(\mathbf{r}, t) = -\frac{iq}{\pi v^2} \int_{-\infty}^{\infty} \frac{1}{\epsilon(\omega)} K_0\left(\frac{|\omega|}{v} r\right) \exp\left[i\omega\left(\frac{z}{v} - t\right)\right] \omega \, d\omega,$$

$$B(\mathbf{r}, t) = \frac{q}{\pi c} \int_{-\infty}^{\infty} k K_1(kr) \exp\left[i\omega\left(\frac{z}{v} - t\right)\right] d\omega, \tag{8.16}$$

$$k = \frac{|\omega|}{c}\left[1 - \frac{v^2}{c^2}\epsilon(\omega)\right]^{1/2}.$$

Substituting (8.16) into (8.15), and noting that $\epsilon(\omega) = 1 - \Omega^2/\omega^2$ when spatial dispersion is neglected, we find as a result of integration

$$\frac{\partial E}{\partial t} = -\frac{q^2 \Omega^3 b_0}{vc} K_0\left(\frac{\Omega b_0}{v}\right) K_1\left(\frac{\Omega b_0}{e}\right). \tag{8.17}$$

Putting $(\Omega/v)b_0 \ll 1$, and using the asymptotic expressions for the MacDonald functions $(K_0(x) \simeq \ln 2/\gamma x,\ \gamma = 1.78107...$ and $K_1(x) \simeq 1/x$ for $x \ll 1)$, we obtain

$$\frac{\partial E}{\partial t} = -\frac{4\pi n_0 e^2 q^2}{mv} \ln\frac{2}{\gamma} \frac{v}{\Omega b_0}. \tag{8.18}$$

Comparing with (8.13) we see that

$$k_0 = \frac{2}{\gamma}\frac{1}{b_0}, \qquad \frac{2}{\gamma} = 1.123... \tag{8.19}$$

In order to determine the energy losses due to the short-range collisions (domain of large k), the collisions of the particle with the individual electrons of the plasma should be considered. If n_0 is the density of the plasma electrons and b the collision parameter, the mean number of collisions within the time dt for which the quantity b will be between b and $b + db$, is $2\pi n_0 vb\, db\, dt$. The energy lost by a particle during a collision with a plasma electron equals

$$\Delta E = -\frac{2e^2q^2}{mv^2}\frac{1}{b^2 + \xi^2}, \qquad \xi = \frac{eq(m + M)}{Mmv^2}, \qquad (8.20)$$

where q is the charge and M the mass of the particle. Multiplying this expression by $2\pi n_0 vb\, db\, dt$, we find the energy losses due to short-range collisions between the particle and the plasma electrons

$$\frac{\partial E}{\partial t} = -\frac{4\pi n_0 e^2 q^2}{mv}\int_0^{b_0}\frac{db\, b}{b^2 + \xi^2}, \qquad (8.21)$$

where integration with respect to b is up to a certain value b_0, for which the plasma electrons may be considered as free electrons. Assuming that $b_0 \gg \xi$, we obtain

$$\frac{\partial E}{\partial t} = -\frac{4\pi n_0 e^2 q^2}{mv}\ln\frac{Mmv^2}{eq(m + M)}b_0. \qquad (8.22)$$

Combining (8.18) and (8.22), we obtain the following formula for the total energy losses of a moving particle per unit time for $v \gg s$ (1)*

$$\frac{\partial E}{\partial t} = -\frac{4\pi n_0 e^2 q^2}{mv}\ln\frac{2}{\gamma}\frac{Mmv^3}{(m + M)eq\Omega}. \qquad (8.23)$$

This formula may be obtained directly from (8.13) by formally taking $(2/\gamma)[Mmv^2/(m + M)eq]$ as k_0.†

Such an analysis is valid if $[Mmv^2/(m + M)eq] < 1/\lambda$, where λ is the wavelength of the plasma electron in a reference system connected with the moving particle $(\lambda = [\hbar(m + M)]/Mmv)$. Otherwise, quantum-mechanical effects play an essential part.

The condition of applicability of the classical analysis may be written in the form

$$v < eq/\hbar. \qquad (8.24)$$

---

* Formula (8.23) corresponds to Bohr's classical retardation formula (2).
† The reciprocal Debye radius is taken as the maximum value k_0 in Pines and Bohm (3).

We see that for sufficiently high particle velocities this condition is violated. If $v > eq/\hbar$, the upper limit of integration b in (8.13) should be connected with the maximum transfer of momentum of the moving particle to the electron of the medium. For sufficiently high transfers the electrons may be considered free, hence, according to (8.20) the maximum momentum transfer equals

$$\hbar k_0 = \frac{2eq}{b_0 v} . \tag{8.25}$$

Substituting $\chi_0 = 2eq/\hbar b_0 v$ into (8.13) and combining with (8.22), we obtain the total energy losses as [Lindhard (4)]

$$\frac{\partial E}{\partial t} = -\frac{4\pi n_0 e^2 q^2}{mv} \ln \frac{2Mmv^2}{(m+M)\hbar\Omega} . \tag{8.26}$$

Formula (8.26) may be obtained directly from (8.10) if the quantum-mechanical expression is used for the plasma dielectric permittivity ϵ_l and recoil of the particle is taken into account.

Formulas (8.23) and (8.26) determine the total energy losses of a moving particle only if the particle mass considerably exceeds the electron mass. The interaction between the particle and the fluctuating field in the plasma plays an essential role in the motion of a light particle (electron) through a plasma.

4. Change in Energy of a Moving Charge due to Field Fluctuations in the Plasma

Let us consider the change in energy of a charge in motion in a plasma, which is due to electromagnetic field fluctuations (5). The equation of motion of the charged particle is written as

$$\dot{\mathbf{v}}(t) = \frac{q}{M}\mathbf{E}(\mathbf{r}(t), t), \tag{8.27}$$

where $\mathbf{E}(\mathbf{r}(t), t)$ is the electric field at the point where the particle is (q and M are the particle charge and mass). Integrating the motion equation (8.27) we find

$$\mathbf{v}(t) = \mathbf{v}_0 + \frac{q}{M}\int_{t_0}^{t} dt' \, \mathbf{E}(\mathbf{r}(t'), t'), \tag{8.28}$$

$$\mathbf{r}(t) = \mathbf{r}_0 + \mathbf{v}_0(t - t_0) + \frac{q}{M} \int_{t_0}^{t} dt' \int_{t_0}^{t'} dt'' \, \mathbf{E}(\mathbf{r}(t''), t'') \qquad (8.29)$$

($\mathbf{r}_0$ and $\mathbf{v}_0$ are the particle radius–vector and velocity at the intitial time t_0). Evidently, the mean change in energy of the moving particle per unit time equals

$$\frac{\partial E}{\partial t} = q \langle \mathbf{v}(t) \, \mathbf{E}(\mathbf{r}(t), t) \rangle, \qquad (8.30)$$

where the brackets $\langle \cdots \rangle$ denote the statistical averaging operation.

Let us select a sufficiently large time segment Δt as compared with the period of the random fluctuations of the electric field in the plasma, but small as compared with the time during which the particle motion changes in an essential manner. Since the particle trajectory differs only slightly from a straight line during this time segment, the particle velocity and the field acting on the particle at time $t = t_0 + \Delta t$ may be represented approximately as

$$\mathbf{v}(t) \simeq \mathbf{v}_0 + \frac{q}{M} \int_{t_0}^{t} dt' \, \mathbf{E}(\mathbf{r}_0(t'), t'),$$

$$\qquad (8.31)$$

$$\mathbf{E}(\mathbf{r}(t), t) \simeq \mathbf{E}(\mathbf{r}_0(t), t) + \frac{q}{M} \int_{t_0}^{t} dt' \int_{t_0}^{t'} dt'' E_j(\mathbf{r}_0(t''), t'') \frac{\partial}{\partial x_{0j}} \mathbf{E}(\mathbf{r}_0(t), t),$$

where $\mathbf{r}_0(t)$ is the particle radius–vector during uniform and straight-line motion

$$\mathbf{r}_0(t) = \mathbf{r}_0 + \mathbf{v}_0(t - t_0). \qquad (8.32)$$

Substituting (8.31) into (8.30), we obtain the following expression for the mean change in energy of the moving particle per unit time, to the accuracy of quadratic terms in q:

$$\frac{\partial E}{\partial t} = q \langle \mathbf{v}_0 \mathbf{E}(\mathbf{r}_0(t), t) \rangle$$

$$+ \frac{q^2}{M} \left\langle \int_{t_0}^{t} dt' \int_{t_0}^{t'} dt'' \, \mathbf{E}(\mathbf{r}_0(t''), t'') \frac{\partial}{\partial x_{0j}} \mathbf{v}_0 \mathbf{E}(\mathbf{r}_0(t), t) \right\rangle$$

$$+ \frac{q^2}{M} \left\langle \int_{t_0}^{t} dt' \, \mathbf{E}(\mathbf{r}_0(t'), t') \, \mathbf{E}(\mathbf{r}_0(t), t) \right\rangle. \qquad (8.33)$$

Evidently, the mean value of the fluctuating part of the field equals zero, hence $\langle \mathbf{E}(\mathbf{r}, t) \rangle$ agrees with the intensity of the field produced by the particle itself in the plasma. Therefore, the first member in (8.33) characterizes the customary polarization energy losses of the moving particle and is determined by (8.11):

$$\frac{\partial E}{\partial t}\bigg|_{\text{pol}} = \frac{q^2}{2\pi^2} \text{Im} \int dk \, \frac{\omega}{k^2 \epsilon_l(\omega k)}, \qquad \omega = \mathbf{k} \mathbf{v}_0. \qquad (8.34)$$

Let us note that the first member in (8.33), as well as the second and third, is proportional to the square of the charge of the moving particle since the mean value of the field is proportional to q. The second and third members in (8.33) characterize the statistical change in the energy of a moving particle, associated with the fluctuations of both the electric field in the plasma, and the velocity of the particle itself under the effect of this field. The second member in (8.33) determines the dynamic friction of the particle due to the presence of space-time correlations between the fluctuating electric fields in the plasma. The presence of such correlations leads to additional losses (in addition to the polarization losses) in the energy of the moving particle. Inverting the order of integration with respect to dt' and dt'' in the second member of (8.33), integrating with respect to dt' and introducing the new variable $\xi = t - t''$, we obtain

$$\frac{\partial E}{\partial t}\bigg|_{\text{II}} = \frac{q^2}{M} \int_0^{\Delta t} d\xi \, \xi \left\langle \mathbf{E}_j(\mathbf{r}_0(t - \xi), t - \xi) \frac{\partial}{\partial x_{0j}} v_{0i} \mathbf{E}_i(\mathbf{r}_0(t), t) \right\rangle. \qquad (8.35)$$

Since the correlation functions for the field fluctuations is exponentially small for large Δt, the upper limit of integration in (8.35) may be extended to infinity ($\Delta t \to \infty$). Representing the field $E_i(\mathbf{r}, t)$ as space-time Fourier integrals, we find for the statistical energy losses of guided particle motion:

$$\frac{\partial E}{\partial t}\bigg|_{\text{II}} = \frac{q^2}{16\pi^3 M} \int d\mathbf{k}\omega \, \frac{\partial}{\partial \omega} \langle E_l^2 \rangle_{\mathbf{k}\omega}, \qquad \omega = \mathbf{k} \mathbf{v}_0. \qquad (8.36)$$

The third member in (8.33) determines the statistical change in the particle energy associated with the correlations between the fluctuating change in the velocity of the particle itself and the fluctuating electric field in the plasma. The presence of such correlations leads to a statistical magnification of the energy of the moving particle. It is easy to obtain the following expression for the statistical magnification of the particle energy:

$$\frac{\partial E}{\partial t}\Big|_{\text{III}} = \frac{q^2}{16\pi^3 M} \int d\mathbf{k} \langle E_l^2 \rangle_{\mathbf{k}\omega} , \qquad \omega = \mathbf{k}\mathbf{v}_0 . \qquad (8.37)$$

Combining (8.36) and (8.37), we obtain

$$\frac{\partial E}{\partial t}\Big|_{\text{st}} = \frac{q^2}{16\pi^3 M} \int d\mathbf{k} \, \frac{\partial}{\partial \omega} (\omega \langle E_l^2 \rangle_{\mathbf{k}\omega}), \qquad \omega = \mathbf{k}\mathbf{v}_0 \qquad (8.38)$$

for the total statistical change in the energy of the moving particle.

In the case of an isotropic plasma, the correlation function of the longitudinal field $\langle E_l^2 \rangle_{\mathbf{k}\omega}$ depends only on the modulus $\mathbf{k}$. Integrating with respect to the angles in (8.38), the statistical change in the energy of the moving particle may be reduced to

$$\frac{\partial E}{\partial t}\Big|_{\text{st}} = \frac{q^2}{4\pi^2 M v_0^3} \int_0^{k_0 v_0} d\omega \, \omega^2 \langle E_l^2 \rangle_{\omega/v_0,\omega} , \qquad (8.39)$$

where $\mathbf{k}_0$ is the maximum value of the wave vector determining the domain of applicability of the macroscopic analysis.

Since the spectral density of the field fluctuations $\langle E_l^2 \rangle_{\mathbf{k}\omega}$ is positive by definition, then according to (8.39) the particle energy in an isotropic plasma will grow because of the particle interaction with the fluctuating longitudinal field. When charged particles move in an anisotropic plasma (plasma permeated by charged particles; plasma in a magnetic field, etc.), interaction with the fluctuating field may lead to diminution in the particle energy.

In contrast to the polarization energy losses, which are independent of the mass of the moving particle, the statistical change in the energy is inversely proportional to the mass of the moving particle. Hence, statistical effects may be neglected in the motion of a heavy particle, with a mass considerably greater than the electron mass, in a plasma. In the case of the motion of a light particle (electron), the statistical effects will be of the same order of magnitude as the polarization effects.

Let us evaluate the polarization losses and the statistical magnification of the energy of a moving particle in the case of an electron plasma for which the spectral distribution of the field fluctuations is determined by the formula:

$$\langle E_l^2 \rangle_{\mathbf{k}\omega} = \frac{8\pi T}{\omega} \frac{\mathrm{Im}\,\epsilon_l(\omega, \mathbf{k})}{|\,\epsilon_l(\omega, \mathbf{k})|^2}.$$

Utilizing (3.23) for the longitudinal dielectric permittivity and integrating with respect to the absolute value of the wave vector $\mathbf{k}$ between zero and some maximum value k_0 for which the macroscopic analysis is still valid in (8.37) and (8.38), we obtain the following general formulas:

$$\frac{\partial E}{\partial t}\bigg|_{\text{pol}} = -\frac{16\,\sqrt{\pi}n_0\,e^2q^2}{mv_0} \int_0^{\zeta} dz\, z^2\, e^{-z^2} L(z, ak_0), \qquad (8.40)$$

$$\frac{\partial E}{\partial t}\bigg|_{\text{st}} = \frac{8\,\sqrt{\pi}n_0\,e^2q^2}{Mv_0}\, \zeta e^{-\zeta^2} L(\zeta, ak_0), \qquad (8.41)$$

where $\zeta = \sqrt{\tfrac{3}{2}}\,v_0/s$ and

$$L(z, ak_0) = \ln ak_0 - \frac{1}{4}\ln\{[1 - \varphi(z)]^2 + \pi z^2\, e^{-2z^2}\}$$

$$- \frac{1}{2\sqrt{\pi}}\frac{1 - \varphi(z)}{z\, e^{-z^2}} \left\{\frac{\pi}{2} - \arctan\frac{1 - \varphi(z)}{\sqrt{\pi}z\, e^{-z^2}}\right\}. \qquad (8.42)$$

(The reciprocal of the minimum parameter of long-range collisions $k_0 \sim Mmv^2/[(M + m)eq]$ may be taken as k_0.) For sufficiently high temperature and low electron density in the plasma $ak_0 \gg 1$. Limiting ourselves in (8.40) and (8.41) to taking account of the first main terms $(\sim \ln ak_0)$, we find*

$$\frac{\partial E}{\partial t}\bigg|_{\text{pol}} = -\frac{4\pi n_0 e^2 q^2}{ms}\sqrt{\frac{3}{2}}\frac{\Phi(\zeta) - \zeta\Phi'(\zeta)}{\zeta}\ln ak_0, \qquad (8.43)$$

$$\frac{\partial E}{\partial t}\bigg|_{\text{st}} = \frac{4\pi n_0 e^2 q^2}{Ms}\sqrt{\frac{3}{2}}\frac{2}{\sqrt{\pi}}e^{-\zeta^2}\ln ak_0, \qquad (8.44)$$

* Kalman and Ron (6) obtained analogous results. Tsytovich (7) considered the question of the passage of charged particles through a plasma when radiation is present in the plasma.

where $\Phi(\zeta)$ is the customary error function

$$\Phi(\zeta) = \frac{2}{\sqrt{\pi}} \int_0^\zeta dz \, e^{-z^2}.$$

In the limiting cases of low and high particle velocities, approximate expressions

$$\frac{\partial E}{\partial t}\bigg|_{\text{pol}} = -\frac{4\pi n_0 e^2 q^2}{ms} \sqrt{\frac{6}{\pi}} \frac{v_0^2}{s^2} \ln ak_0, \quad v_0 \ll s, \tag{8.45}$$

$$\frac{\partial E}{\partial t}\bigg|_{\text{pol}} = -\frac{4\pi n_0 e^2 q^2}{mv_0} \ln ak_0, \quad v_0 \gg s \tag{8.46}$$

may be obtained for the polarization energy losses.

The polarization losses, statistical magnification, and total energy change per unit time [in the units $(4\pi n_0 e^2 q^2/ms) \sqrt{\frac{3}{2}} \ln ak_0$] are presented in Fig. 16 as a function of the velocity for an electron moving in a plasma $(M = m)$.

As we see, in the case of particle motion in an equilibrium plasma

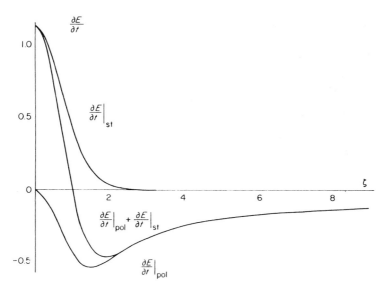

FIG. 16.

at a velocity considerably exceeding the mean thermal electron velocity ($v_0 \gg s$), the influence of the fluctuating field may be neglected. Hence, the total change in particle energy reduces to polarization losses in the energy of the directed motion. If the velocity of the charged particle motion is comparable to the thermal electron velocity, the role of the fluctuating interaction then becomes significant. In particular, if the particle velocity is less than the thermal velocity of the plasma electrons, the energy of the particles moving in the plasma increases because of the fluctuating interaction.

Let us note that (8.43) and (8.44) are a good approximation if the large quantity $ak_0 \gg 1$ is in the logarithm. This condition is satisfied if the energy of particle interaction in the plasma is small as compared with the kinetic energy. Comparing (8.46) with (8.13), we see that (8.13) differs from (8.34) by the factor v_0/s in the logarithm. This indicates that taking account of just the principal term in (8.42) is not a good enough approximation at high particle velocities.

5. Influence of Ions on the Energy Change of a Particle Moving in a Plasma

Let us examine how taking account of the ion motion in a plasma affects the change in energy of a moving charged particle (5). Using (5.52) for the spectral distribution of the fluctuations and (5.8) for the dielectric permittivity, the following formulas for the polarization losses and the statistical magnification of the energy of a moving particle in a nonisothermal electron–ion plasma may be obtained on the basis of (8.34) and (8.38):

$$\frac{\partial E}{\partial t}\bigg|_{\text{pol}} = -\frac{4\pi n_0 e^2 q^2}{ms}\sqrt{\frac{3}{2}}\left\{\frac{\Phi(\zeta) - \zeta\Phi'(\zeta)}{\zeta} + \frac{t}{\mu}\frac{\Phi(\mu\zeta) - \mu\zeta\Phi'(\mu\zeta)}{\mu\zeta}\right\}\ln ak_0 ,$$

(8.47)

$$\frac{\partial E}{\partial t}\bigg|_{\text{st}} = \frac{4\pi n_0 e^2 q^2}{Ms}\sqrt{\frac{3}{2}}\frac{2}{\sqrt{\pi}}\{e^{-\zeta^2} + \mu\, e^{-\mu^2\zeta^2}\}\ln ak_0 .$$

(8.48)

Here $ak_0 \gg 1$, $t = T_e/T_i$ and $\mu^2 = t(M_i/m)$.

If the velocity of the charged particle is considerably greater than the thermal electron velocity $v_0 \gg s$, the change in energy of the moving particle is due primarily to particle interaction with the electrons. The interaction between the charged particle and the ions may be neglected here and (8.47) and (8.48) go over into (8.43) and (8.44).

As the particle velocity diminishes $v_0 < s$ its interaction with the plasma electrons diminishes relatively, and its interaction with the plasma ions becomes essential. If the condition $1 \gg \zeta \gg \mu^{-1}$ is satisfied, the polarization energy losses are determined by the expression

$$\frac{\partial E}{\partial t}\bigg|_{pol} = -\frac{4\pi n_0 e^2 q^2}{ms} \sqrt{\frac{3}{2}} \left\{ \frac{4}{3\sqrt{\pi}} \zeta^2 + \frac{m}{M_i} \frac{1}{\zeta} \right\} \ln ak_0 . \tag{8.49}$$

In particular, for $\zeta^3 \gg m/M_i$ we have

$$\frac{\partial E}{\partial t}\bigg|_{pol} = -\frac{4\pi n_0 e^2 q^2}{M_i v_0} \ln ak_0 . \tag{8.50}$$

For $\zeta \ll \mu^{-1}$ we find for the polarization losses

$$\frac{\partial E}{\partial t}\bigg|_{pol} = -\frac{4\pi n_0 e^2 q^2}{ms} \sqrt{\frac{6}{\pi}} \frac{v_0^2}{s^2} t\mu \ln ak_0 . \tag{8.51}$$

In this case the statistical change in energy equals

$$\frac{\partial E}{\partial t}\bigg|_{st} = \frac{4\pi n_0 e^2 q^2}{ms} \sqrt{\frac{6}{\pi}} \mu \ln ak_0 . \tag{8.52}$$

Hence, at low velocities the change in energy of a particle moving in a plasma is determined primarily by particle interaction with the ions.

6. Energy Losses of a Charged Particle in a Nonequilibrium Plasma

When a charged particle moves in a nonequilibrium plasma (especially if the plasma is near the boundary of the stability domain), the change in particle energy is determined mainly by the interaction with the fluctuating field. As an example, let us consider a plasma through which a neutral beam of charged particles passes. If the beam density is sufficiently small, the polarization energy losses of a particle moving through the plasma do not change substantially. Interaction of the charged particle with the fluctuating longitudinal field will lead to additional energy losses associated with the excitation of Langmuir oscillations in the plasma.

Utilizing (8.38) and (5.96), the following expression may easily be found for the particle energy losses:

$$\frac{\partial E}{\partial t} = -\frac{q^2 \Omega^2}{\pi M} \int d\omega \, d\mathbf{k} \, \tilde{T} \, \delta(\omega^2 - \Omega^2) \, \delta'(\omega - \mathbf{kv}), \qquad (8.53)$$

$$\tilde{T} = T' \Big/ \Big| 1 - \frac{u}{\tilde{u}} \cos \vartheta \Big|, \qquad \tilde{u} = \frac{\omega}{k}.$$

In the simplest case, when $\mathbf{v} \parallel \mathbf{u}$ we find by integrating in (8.53)

$$\frac{\partial E}{\partial t} = -\frac{q^2 T'}{2Mv^2} \frac{uk_0^2}{|1 - (u/v)|^2}, \qquad v > u. \qquad (8.54)$$

Formulas (8.53) and (8.54) describe the charged-particle interaction with fluctuating Langmuir waves in a plasma. According to (8.54) the intensity of Langmuir wave excitation by the moving charge grows anomalously if the charged particle velocity v approaches the beam velocity u. The magnitude of the energy losses due to particle interaction with the fluctuating field may hence considerably exceed the losses associated with binary collisions.

REFERENCES

1. A. I. Akhiezer and A. G. Sitenko, ZETF **23**, 161 (1952).
2. N. Bohr, Dan. Mat. Fys. Medd. **18**(8) (1948).
3. D. Pines and D. Bohm, Phys. Rev. **85**, 338 (1952).
4. J. Lindhard, Dan. Mat. Fys. Medd. **28**(8) (1954).
5. A. G. Sitenko and Tsien Yu-Tai, ZTF **32**, 1325 (1962) (Sov. Phys.—Tech. Phys.).
6. G. Kalman and A. Ron, Ann. Phys. **16**, 118 (1961).
7. V. N. Tsytovich, ZETF **42**, 803 (1962); **44**, 846 (1963) (Sov. Phys. JETP); Izv. Vuz, Radiofizika **5**, 1078 (1962) (CFSTI).

9
Dynamic Friction and
Diffusion Coefficients in a Plasma

1. Fokker–Planck Equation

Slow irreversible processes for which the relaxation time considerably exceeds the time of the particle mean-free-path are possible in a plasma in the nonequilibrium state. The characteristic peculiarities of such irreversible processes in the plasma are due to the far-reaching character of the Coulomb interaction between charged particles.

Long-range collisions, for which deflections of the colliding particles occur only at low angles and with a small change in velocity, play an essential part in the irreversible processes because of the far-reaching character of the Coulomb forces. Hence, to describe such processes it is possible to use the Fokker–Planck equation in which the effect of the collisions reduces to particle diffusion into the velocity space.

The Fokker–Planck equation expressing the law of conservation of the number of particles determines the change in the distribution function of an isolated group of particles f per unit time due to the diffusion of particles into the velocity space:

$$\frac{\partial f}{\partial t} = - \frac{\partial}{\partial v_i}(D_i f) + \frac{1}{2}\frac{\partial^2}{\partial v_i\,\partial v_j}(D_{ij}f), \qquad (9.1)$$

where D_i is the dynamic friction coefficient defined as the mean change in the particle velocity per unit time

$$D_i = \frac{\langle \Delta v_i \rangle}{\Delta t} \tag{9.2}$$

and D_{ij} is the tensor of the diffusion coefficients defined according to the equality

$$D_{ij} = \frac{\langle \Delta v_i \, \Delta v_j \rangle}{\Delta t} \, . \tag{9.3}$$

The derivations of (9.1) is based on the assumption that small changes in the velocity Δv_i are most probable, and that powers of Δv_i higher than the second may be neglected.

Terms in the Fokker–Planck equation containing higher powers of Δv_i will correspond to taking account of short-range collisions, at which the particle is scattered at large angles. However, such scattering may not be considered as diffusion into the velocity space. Subsequent terms in the Fokker–Planck equation may be evaluated on the basis of the theory of binary collisions. As may be shown, summation of these terms leads to the Boltzmann collision integral. In the general case, the friction coefficient D_i and the diffusion coefficients D_{ij} depend on both the distribution function of the isolated group of particles (test particles), and on the distribution function of the remaining particles of the plasma (field particles). Considering the motion of the individual test particle in the plasma and assuming that the remaining field particles are in equilibrium, the friction and diffusion coefficients may be evaluated explicitly.

Landau (1) first derived the kinetic equation of type (9.1) taking into account the Coulomb collisions between charged particles. Chandrasekhar (2) and Spitzer *et al.* (3, 4) also calculated the friction and diffusion coefficients for an isotropic plasma without taking account of ion motion. Only long-range collisions with selective parameters less than the Debye radius in the plasma are hence taken into account. The results (2, 3) may be obtained on the basis of a macroscopic analysis without introduction of an arbitrary cutoff parameter, if the friction and diffusion coefficients are expressed in terms of the spectral distribution of the fluctuations of the longitudinal electric field in the plasma.*

Let us consider the motion of some test particle in a plasma, where its charge and mass will be denoted by q and M. The equation of particle motion is

$$\dot{v}_i(t) = \frac{q}{M} E_i(\mathbf{r}(t), t). \tag{9.4}$$

Formal integration of this equation yields (8.29). Selecting the time segment Δt sufficiently large as compared with the period of random field fluctuations in the plasma, but small as compared with the time during which the motion of the particle changes the equation of motion in an essential manner, (9.4) may be represented approximately as

$$\dot{v}_i(t) = \frac{q}{M} E_i(\mathbf{r}_0(t), t) + \frac{q^2}{M^2} \int_{t_0}^{t} dt' \int_{t_0}^{t'} dt'' \, E_j(\mathbf{r}_0(t''), t'') \frac{\partial}{\partial x_{0j}} E_i(\mathbf{r}_0(t), t),$$
$$\tag{9.5}$$

$$\mathbf{r}_0(t) = \mathbf{r}_0 + \mathbf{v}(t - t_0), \quad t = t_0 + \Delta t.$$

Averaging (9.5) with respect to the fluctuations, we indeed find the coefficient of dynamic friction.

Evidently, only the component of D_i in the direction of the velocity of the sample particle v_i, is not zero in the case of an isotropic plasma, hence $D_i = (v_i/v)D$. The quantity D characterizes the mean change in the absolute value of the test-particle velocity per unit time. It is easy to see that D is directly connected with the energy losses of the guided particle motion per unit time $\partial E/\partial t$ by means of the relationship

$$D = \frac{1}{Mv} \frac{\partial E}{\partial t}.$$

Using (8.34) and (8.36) we find

$$D = \frac{q^2}{2\pi^2 Mv} \operatorname{Im} \int d\mathbf{k} \frac{\omega}{k^2 \epsilon_l(\omega, \mathbf{k})} + \frac{q^2}{16\pi^3 M^2 v} \int d\mathbf{k}\omega \frac{\partial}{\partial \omega} \langle E_l^2 \rangle_{\mathbf{k}\omega}, \tag{9.6}$$

$$\omega = \mathbf{k}\mathbf{v}.$$

* Thompson and Hubbard (5) proposed such a method of determining the diffusion coefficients. See also References 6 and 7. We will follow the method in Reference 6.

The first member in (9.6) determines the dynamic friction of the particle due to its interaction with the electric field originating in the plasma from the motion of the particle itself. This interaction leads to the customary polarization energy losses of the particle and may be called polarization friction.

The second member in (9.6) determines the additional dynamic friction of the particle associated with the presence of space-time correlations between the fluctuating electric fields in the plasma. As we have seen, the presence of such correlations leads to additional energy losses of the guided particle motion. Let us note that the dynamic friction of the particle is associated only with the fluctuations of the longitudinal electric field in the plasma.

In an analogous manner it is easy to obtain a general expression for the tensor of the diffusion coefficients D_{ij}, which are determined completely by the spectral distribution of the longitudinal electric-field fluctuations in the plasma, exactly as is the dynamic friction coefficient D in the nonrelativistic case. Since the diffusion coefficients are purely of fluctuating origin, it is sufficient to use the equation of motion of the test particle (9.5) in finding the D_{ij}, wherein only terms linear in the field are taken into account:

$$D_{ij} = \langle \dot{v}_i \Delta v_j \rangle + \langle \Delta v_i \dot{v}_j \rangle = \frac{q^2}{M^2} \int_0^{\Delta t} d\xi \, \langle E_j(\mathbf{r}_0(t-\xi), t-\xi) \, E_i(\mathbf{r}_0(t), t) \rangle. \tag{9.7}$$

Let Δt tend to infinity and transforming to Fourier components, we finally obtain

$$D_{ij} = \frac{q^2}{8\pi^3 M^2} \int d\mathbf{k} \, \frac{k_i k_j}{k^2} \langle E_l^2 \rangle_{\mathbf{k}\omega}, \qquad \omega = \mathbf{k}\mathbf{v}. \tag{9.8}$$

In an isotropic plasma the tensor of the diffusion coefficients may be represented in the form

$$D_{ij} = \frac{v_i v_j}{v^2} D_\parallel + \frac{1}{2}\left(\delta_{ij} - \frac{v_i v_j}{v^2}\right) D_\perp, \tag{9.9}$$

$$D_\parallel = \frac{q^2}{8\pi^3 M^2 v^2} \int d\mathbf{k} \, \frac{\omega^2}{k^2} \langle E_l^2 \rangle_{\mathbf{k}\omega}, \tag{9.10}$$

$$D_\perp = \frac{q^2}{8\pi^3 M^2} \int d\mathbf{k} \left(1 - \frac{\omega^2}{k^2 v^2}\right) \langle E_l^2 \rangle_{\mathbf{k}\omega}, \qquad \omega = \mathbf{k}\mathbf{v}. \tag{9.11}$$

The longitudinal component of diffusion $D_{\parallel}$ characterizes the mean change in the square of the velocity component along the direction of particle motion; the transverse diffusion component $D_{\perp}$ characterizes the mean change in the square of the velocity component in the perpendicular direction.

2. Dynamic Friction and Diffusion Coefficients in an Electron Plasma

First let us consider dynamic friction and diffusion in an equilibrium electron plasma neglecting ion motion. The dynamic friction coefficient in a plasma is determined by the general formula (9.6). Utilizing (3.23) and (4.17) for the longitudinal dielectric permittivity and the spectral distribution of the field fluctuations, we find as a result of integrating (9.6)

$$D = D_{\text{pol}} + D_{\text{fl}} = -\frac{6}{\sqrt{\pi}}\frac{q^2\Omega^2}{Ms^2}\left(1 + \frac{m}{M}\right)\zeta^{-2}\int_0^{\zeta} dz\, z^2\, e^{-z^2}L(z, ak_0),$$

$$(9.12)$$

where $\zeta = \sqrt{\frac{3}{2}}\,(v/s)$ and $L(z, ak_0)$ is determined by (8.42). The one in the parentheses of (9.12) corresponds to the polarization friction coefficient D_{pol}, and the component m/M to the coefficient of fluctuating friction D_{fl}. For a heavy test particle ($M \gg m$) the polarization friction D_{pol} is considerably greater than the fluctuating friction D_{fl}. If an electron is taken as test particle, then $D_{\text{pol}} = D_{\text{fl}}$.

Let us be limited in the most interesting case of high temperatures and low electron density ($ak_0 \gg 1$), to taking only the fundamental term $L(z, ak_0) \simeq \ln ak_0$ into account in (8.42), then we obtain the known Chandrasekar formula from (9.12):

$$D = -\frac{3q^2\Omega^2}{Ms^2}\left(1 + \frac{m}{M}\right)G(\zeta)\ln ak_0.\qquad(9.13)$$

The function $G(\zeta)$ is expressed in terms of the error function $\Phi(\zeta)$ by using the formula

$$G(\zeta) = \frac{\Phi(\zeta) - \zeta\Phi'(\zeta)}{2\zeta^2}.\qquad(9.14)$$

In the limiting case of low and high velocities of the moving particle we have

$$D = -\frac{2}{\sqrt{\pi}}\frac{q^2\Omega^2}{Ms^2}\left(1 + \frac{m}{M}\right)\zeta \ln ak_0, \quad \zeta \ll 1, \tag{9.15}$$

$$D = -\frac{3}{2}\frac{q^2\Omega^2}{Ms^2}\left(1 + \frac{m}{M}\right)\frac{1}{\zeta^2} \ln ak_0, \quad \zeta \gg 1. \tag{9.16}$$

For slow particles in a plasma the dynamic friction increases linearly as the particle velocity increases, while for fast particles the friction decreases in inverse proportion to the velocity squared.

The maximum of the dynamic friction force is determined from the condition

$$\frac{d^2}{d\zeta^2}\left(\frac{\Phi(\zeta)}{\zeta}\right) = 0. \tag{9.17}$$

This condition is satisfied for $\zeta \simeq 0.97$. Hence, the friction will be greatest when the velocity of the particle moving in the vacuum almost agrees with the mean thermal velocity of the plasma electrons. In exactly the same manner we find for the longitudinal and transverse diffusion coefficients,

$$D_{\parallel} = \frac{4}{\sqrt{\pi}}\frac{q^2\Omega^2}{M^2v}m\zeta^{-2}\int_0^\zeta dz\, z^2\, e^{-z^2}L(z, ak_0), \tag{9.18}$$

$$D_{\perp} = \frac{4}{\sqrt{\pi}}\frac{q^2\Omega^2}{M^2v}m\zeta^{-2}\int_0^\zeta dz(\zeta^2 - z^2)\, e^{-z^2}L(z, ak_0). \tag{9.19}$$

Comparing (9.18) and (9.19) with (9.12), we see that the diffusion coefficients $D_{\parallel}$ and $D_{\perp}$ contain an excess factor M in the denominator, as compared with the dynamic friction coefficient D. Hence, for heavy test particles ($M \gg m$) the retardation due to the dynamic friction is so much more important than the dispersion in particle velocity associated with diffusion.

Retaining only the fundamental terms in (9.18) and (9.19) we obtain the Chandrasekhar formulas:

$$D_{\parallel} = \frac{2q^2\Omega^2m}{M^2v}\,G(\zeta)\ln ak_0, \tag{9.20}$$

$$D_{\perp} = \frac{2q^2\Omega^2m}{M^2v}\{\Phi(\zeta) - G(\zeta)\}\ln ak_0. \tag{9.21}$$

If the velocity of the moving particle is low ($v \ll s$), the difference between the displacements parallel and perpendicular to the direction of motion vanishes; hence

$$D_\perp = 2D_\parallel, \quad \zeta \ll 1. \tag{9.22}$$

In the case of high-particle velocities ($v \gg s$), the relationship

$$D_\perp = 2\zeta^2 D_\parallel, \quad \zeta \gg 1 \tag{9.23}$$

holds. Hence, if the velocity of the test particles is greater than the mean thermal velocity of the plasma electrons, the diffusion in the velocity space will be primarily lateral, i.e., perpendicular to the original velocity.

3. Friction and Diffusion Coefficients in a Two-Temperature Plasma

The general formulas obtained in the first section also permit the investigation of the friction and diffusion in a nonequilibrium plasma, if only the spectral distribution of the field fluctuations is known. In particular, the dynamic friction and diffusion coefficients in a two-temperature electron–ion plasma may be found by using formulas (9.6), (9.10), and (9.11) (6).

The dielectric permittivity and the spectral distribution of the fluctuations of the longitudinal electric field are determined for a two-temperature plasma by (5.8) and (5.52). Using (9.6), we obtain the following formulas for the dynamic friction coefficients in a two-temperature plasma D_{pol} and D_{fl}:

$$D_{\text{pol}} = -\frac{6}{\sqrt{\pi}} \frac{q^2 \Omega^2}{M s^2} \zeta^{-2} \int_0^\zeta dz \, z^2 (e^{-z^2} + t\mu e^{-\mu^2 z^2}) L(z, t, \mu, ak_0), \tag{9.24}$$

$$D_{\text{fl}} = -\frac{6}{\sqrt{\pi}} \frac{q^2 \Omega^2}{M s^2} \frac{m}{M} \zeta^{-2} \int_0^\zeta dz \, z^2 (e^{-z^2} + \mu^3 e^{-\mu^2 z^2}) L(z, t, \mu, ak_0), \tag{9.25}$$

where

$$L(z, t, \mu, ak_0) = \ln ak_0 - \frac{1}{4} \ln \{[1 - \varphi(z) + t(1 - \varphi(\mu z))]^2$$

$$+ \pi z^2 (e^{-z^2} + t\mu e^{-\mu^2 z^2})^2\}$$

$$- \frac{1}{2\sqrt{\pi}} \frac{1 - \varphi(z) + t(1 - \varphi(\mu z))}{z(e^{-z^2} + t\mu e^{-\mu^2 z^2})}$$

$$\times \left\{\frac{\pi}{2} - \arctan \frac{1 - \varphi(z) + t(1 - \varphi(\mu z))}{\sqrt{\pi} z(e^{-z^2} + t\mu e^{-\mu^2 z^2})}\right\}. \qquad (9.26)$$

Limiting ourselves to taking account of terms with $ak_0 \gg 1$, we obtain approximate formulas for the dynamic friction coefficients:

$$D_{pol} = -\frac{3q^2 \Omega^2}{Ms^2} \{G(\zeta) + tG(\mu\zeta)\} \ln ak_0, \qquad (9.27)$$

$$D_{fl} = -\frac{3q^2 \Omega^2}{Ms^2} \frac{m}{M} \{G(\zeta) + \mu^2 G(\mu\zeta)\} \ln ak_0. \qquad (9.28)$$

Taking account of the ion motion in the plasma leads to an essential magnification of the part of the fluctuating friction. Thus if an electron is taken as test particle, the fluctuating friction may even exceed the polarization friction significantly.

The dependence of the dynamic friction coefficients D_{pol} and D_{fl} on the velocity ζ is presented in Figs. 17 and 18. The dependence of the total dynamic friction coefficient D and D' on the velocity ζ is shown in Fig. 19, if an electron or proton is taken as

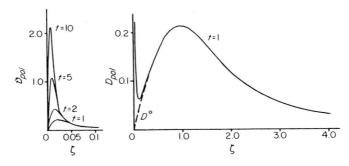

FIG. 17.

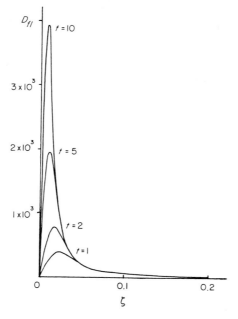

FIG. 18.

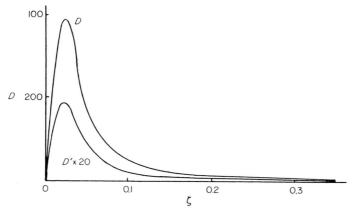

FIG. 19.

test particle. Let us note that for low velocities of the test particles both the polarization friction D_{pol}, and the fluctuating friction D_{fl} are determined mainly by particle interaction with the plasma ions. In an analogous manner it is possible to find the diffusion coefficients $D_\parallel$ and $D_\perp$ for a two-temperature plasma

$$D_\parallel = \frac{4}{\sqrt{\pi}} \frac{q^2\Omega^2 m}{M^2 v} \zeta^{-2} \int_0^\zeta dz\, z^2 (e^{-z^2} + \mu\, e^{-\mu^2 z^2}) L(z, t, \mu, ak_0), \qquad (9.29)$$

$$D_\perp = \frac{q^2\Omega^2 m}{M^2 v} \zeta^{-2} \int_0^\zeta dz\, (\zeta^2 - z^2)(e^{-z^2} + \mu\, e^{-\mu^2 z^2}) L(z, t, \mu, ak_0). \qquad (9.30)$$

In the limiting case $ak_0 \gg 1$ we have

$$D_\parallel = \frac{2q^2\Omega^2 m}{M^2 v} [G(\zeta) + G(\mu\zeta)] \ln ak_0, \qquad (9.31)$$

$$D_\perp = \frac{2q^2\Omega^2 m}{M^2 v} [\Phi(\zeta) - G(\zeta) + \Phi(\mu\zeta) - G(\mu\zeta)] \ln ak_0. \qquad (9.32)$$

It follows from (9.32) that the transverse diffusion depends on the ion motion in an essential way. For $v \gg s$ the transverse diffusion is due, for example, equally to the test-particle interaction with both the electrons and the ions of the plasma. For $s \gg v \gg s_i$ the transverse diffusion is determined primarily by interaction with the ions.

4. Relaxation Time of a Nonisothermal Plasma

By using the expressions found for the friction and diffusion coefficients it is possible to evaluate the relaxation time τ, i.e., the time during which a uniform energy distribution is built up between the electrons and ions in the plasma.

Let us define the plasma relaxation time τ by means of the relationship:

$$\frac{dT_e}{dt} = -\frac{T_e - T_i}{\tau} \qquad (9.33)$$

(T_e and T_i are the temperatures of the plasma electrons and ions).

The change in energy of the test particle per unit time, due to interaction with the plasma particles equals

$$\frac{\partial E}{\partial t} = \frac{M}{2}(2vD + D_\parallel + D_\perp). \tag{9.34}$$

Selecting the electron as test particle, we have

$$\frac{\partial E^e}{\partial t} = \frac{e^2\Omega^2}{v}\left\{-\Phi(\zeta) + 2\zeta\Phi'(\zeta) - \frac{t}{\mu^2}\Phi(\mu\zeta) + \left(1 + \frac{t}{\mu^2}\right)\mu\zeta\Phi'(\mu\zeta)\right\}. \tag{9.35}$$

This expression, averaged with respect to the Maxwell distribution of electron velocities in the plasma determines the change in electron temperature with time

$$\left\langle\frac{\partial E_e}{\partial t}\right\rangle = \frac{3}{2}\frac{dT_e}{dt}. \tag{9.36}$$

The first and second members in (9.35) determine the change in electron energy due to interaction with the remaining electrons in the plasma, and the third and fourth members, to the change in energy due to interaction with the plasma ions. It is easy to verify that the sum of the two first members vanishes after having been averaged since the plasma electrons are in thermal equilibrium with each other. The mean change in energy due to electron interaction with the ions equals

$$\left\langle\frac{\partial E_e}{\partial t}\right\rangle = -\frac{4\sqrt{2\pi}n_0e^4}{m_em_i}\frac{T_e - T_i}{[(T_e/m_e) + (T_i/m_i)]^{3/2}}\ln ak_0. \tag{9.37}$$

Using (9.33) and (9.36), we therefore obtain the known expression for the relaxation time of a nonisothermal plasma [Landau (1), Spitzer (4)]

$$\tau = \frac{3m_em_i}{8\sqrt{2\pi}n_0e^4\ln ak_0}\left(\frac{T_e}{m_e} + \frac{T_i}{m_i}\right)^{3/2}. \tag{9.38}$$

If the distribution functions of the electrons and ions differ from the Maxwellian, equilibrium is first established within the electron and ion gases. Hence, in the case of a strongly nonisothermal plasma the relaxation times determining the isotropization of the distribution functions depend essentially on the long-range interaction between particles associated with the slow waves in a plasma. As has been shown by Ramazashvili *et al.* (8), taking account of particle interaction with the waves leads to a radical decrease in the time of isotropization of the electrons and ions. The relaxation times of isotropic distribution functions to Maxwellian distributions are determined primarily by short-range collisions between particles.

REFERENCES

1. L. D. Landau, *ZETF* **7**, 203 (1937).
2. S. Chandrasekhar, "Principles of Stellar Dynamics." Dover, New York, 1943.
3. R. Cohen, L. Spitzer, and P. Routly, *Phys. Rev.* **80**, 230 (1950).
4. L. Spitzer, "Physics of Fully Ionized Gases," 2nd ed. Wiley (Interscience), New York, 1962.
5. W. Thompson and J. Hubbard, *Rev. Modern Phys.* **32**, 714 (1960).
6. A. G. Sitenko and Tsien Yu-Tai, *ZTF* **32**, 1325 (1962) *(Sov. Phys.-Tech. Phys.)*.
7. J. Hubbard, *Proc. Roy. Soc. (London)* **A260**, 114 (1961); **A261**, 371 (1961).
8. R. R. Ramazashvili, A. A. Rukhadze, and V. P. Silin, *ZETF* **43**, 1323 (1962) *(Sov. Phys. JETP)*.

10

Electromagnetic Wave Scattering
by Fluctuations in Plasma

1. Wave Scattering and Transformation

Let us turn now to the examination of the scattering of electromagnetic waves which occurs when they are propagated in plasma. This scattering is due to thermal fluctuations in the plasma which affect the dielectric properties of the plasma.[*]

As we have seen, various kinds of waves exist in a plasma, which turn out to be mutually independent in a linear approximation. (For example, in the absence of an external magnetic field the waves are separated into longitudinal and transverse, etc.) In reality, because of the nonlinearity of the material equations, each of these waves may be scattered or transformed into a wave of another type by interacting with others. We will speak of scattering if the wave originating in the plasma is a wave of the same kind as the incident wave, and of transformation, if the originating wave is a wave of a different type.

Let us note that not all the conceivable scattering and transformation processes may occur because of the restrictions imposed by the energy and momentum conservation laws on the frequency and wave vector of the scattered or transformed waves. Thus, if a wave with frequency ω_0, wave vector $\mathbf{k}_0$, and dispersion law

---

[*] Electromagnetic wave scattering by fluctuations in plasma has been considered in References (1, 2).

$\omega_0 = \omega_0(\mathbf{k}_0)$ is transformed into a wave with frequency ω, wave vector $\mathbf{k}$, and dispersion law $\omega = \omega(\mathbf{k})$ because of scattering by a wave with frequency $\tilde{\omega}$ and wave vector $\mathbf{q}$, the following conservation laws:

$$\omega = \omega_0 + \tilde{\omega}, \qquad (10.1)$$
$$\mathbf{k} = \mathbf{k}_0 + \mathbf{q}$$

should then be satisfied. Utilizing the dispersion law for the wave by which the scattering is done $\tilde{\omega} = \tilde{\omega}(\mathbf{q})$, we therefore obtain the equation

$$\omega(\mathbf{k}) = \omega_0(\mathbf{k}_0) + \tilde{\omega}(\mathbf{k} - \mathbf{k}_0). \qquad (10.2)$$

In order to be able to accomplish the transition it is necessary that (10.2) have real solutions for $\mathbf{k}$. If this condition is not satisfied, the appropriate transition turns out to be forbidden (3). For example, it is easy to see that scattering of any wave by a wave of the same kind is impossible. According to (10.2), we have

$$\omega(\mathbf{k}) = \omega(\mathbf{k}_0) + \omega(\mathbf{k} - \mathbf{k}_0). \qquad (10.3)$$

Setting $\mathbf{k}$ and $\mathbf{k}_0$ successively equal to zero, we see that (10.3) may be satisfied only if $\omega(\mathbf{k}) = 0$. Therefore, wave scattering by a wave of the same kind may not be accomplished.

Electromagnetic waves being propagated in a plasma because of existing thermal fluctuations may be scattered by these fluctuations, and may also be absorbed with the excitation of longitudinal plasma waves.

2. Electromagnetic Wave Propagation in a Plasma

Considering the wavelength to exceed significantly the mean distance between individual particles ($\lambda \gg n_0^{-1/3}$), the plasma may be considered as a continuous medium characterized by a definite dielectric permittivity. Moreover, if the wavelength turns out to be considerably greater than the Debye radius $\lambda \gg a$, then spatial dispersion may be neglected and the plasma may be considered as a conventional dielectric. The dielectric permittivity of the plasma is hence determined completely by the electrons since for frequencies at which electromagnetic wave propagation is possible in a fully ionized plasma, the influence of the ions may be neglected completely because of the great mass.

Let us present an elementary derivation of the fundamental relationships describing wave propagation in a plasma in this case. Using the Maxwell system (2.1), the equation for the field of the electromagnetic wave may be written as

$$\text{rot rot } \mathbf{E} = -\frac{1}{c^2}\frac{\partial^2 \mathbf{E}}{\partial t^2} - \frac{4\pi}{c^2}\frac{\partial \mathbf{j}}{\partial t}, \tag{10.4}$$

where $\mathbf{j} = en\mathbf{v}$ is the current density associated with the motion of the plasma electrons, n and $\mathbf{v}$ are, respectively, the electron density and velocity. Let us use the hydrodynamic equations for the velocity of electrons in the field of the electromagnetic wave:

$$\frac{\partial v}{\partial t} + (\mathbf{v}\nabla)\mathbf{v} = \frac{e}{m}\left(\mathbf{E} + \frac{1}{c}[\mathbf{v}, \mathbf{B}]\right),$$
$$\frac{\partial n}{\partial t} + \text{div } n\mathbf{v} = 0. \tag{10.5}$$

If the electron density may be considered constant $n = n_0$, then by linearizing (10.5) and substituting into (10.4), we have

$$\text{rot rot } \mathbf{E} + \frac{\hat{\epsilon}}{c^2}\frac{\partial^2 E}{dt^2} = 0, \tag{10.6}$$

where $\hat{\epsilon}$ is the operator of the plasma's dielectric permittivity, whose proper values equal

$$\epsilon(\omega) = 1 - \frac{\Omega^2}{\omega^2}, \qquad \Omega^2 = \frac{4\pi e^2 n_0}{m}. \tag{10.7}$$

Hence, in a homogeneous plasma, the propagation of plane monochromatic transverse electromagnetic waves

$$\mathbf{E}_0(\mathbf{r}, t) = \mathbf{E}_0 e^{i\mathbf{k}_0\mathbf{r} - i\omega_0 t}, \qquad \mathbf{k}_0\mathbf{E}_0 = 0, \tag{10.8}$$

whose frequencies and wave vectors are connected by the relationship

$$k_0^2 = \epsilon(\omega_0)\,\omega_0^2/c^2$$

is possible.

3. Electromagnetic Wave Scattering. Differential Scattering Coefficient

The electron density n may experience a deviation from the equilibrium value n_0

$$n = n_0 + \delta n$$

(δn the fluctuations in the electron density), because of thermal fluctuations in the plasma. Hence, the velocity fluctuations $\delta \mathbf{v}$ will be connected to $\delta \mathbf{n}$ by means of the continuity relationship. An additional current defined by the equality

$$\frac{\partial \mathbf{j}}{\partial t} = \frac{\partial}{\partial t} e \delta n \mathbf{v}_0 + e n_0 \left\{ \frac{e}{mc} [\delta \mathbf{v}, \mathbf{B}_0] - (\mathbf{v}_0 \nabla) \delta \mathbf{v} - (\delta \mathbf{v} \nabla) \mathbf{v}_0 \right\}, \qquad (10.9)$$

where $\mathbf{v}_0$ is the velocity acquired by the electron under the effect of the field of the incident electromagnetic wave

$$\mathbf{v}_0 = i \frac{e}{m \omega_0} \mathbf{E}_0 e^{i \mathbf{k}_0 \mathbf{r} - i \omega_0 t}$$

will originate during propagation of the electromagnetic wave (10.8) in the plasma. This current will also evidently lead to the appearance of scattered electromagnetic waves with altered frequencies and propagation directions.

The field of a scattered electromagnetic wave is determined by

$$\text{rot rot } \mathbf{E} + \frac{\hat{\epsilon}}{c^2} \frac{\partial^2 \mathbf{E}}{\partial t^2} = - \frac{4\pi}{c^2} \frac{\partial \mathbf{j}}{\partial t}. \qquad (10.10)$$

Let us note that the scattered wave $\mathbf{E}$ may contain both transverse and longitudinal parts, i.e., in addition to the pure scattering whereby a transverse wave with an altered frequency and direction of propagation originates, absorption of the incident wave accompanied by excitation of longitudinal oscillations in the plasma (transformation of waves) is also possible.

Retaining only the transverse part of the current $\mathbf{j}_t$ (div $\mathbf{j}_t = 0$) in (10.10) we obtain the following equation determining the scattered electromagnetic waves:

$$\Delta \mathbf{E} - \frac{\epsilon}{c^2} \frac{\partial^2 E}{\partial t^2} = \frac{4\pi}{c^2} \frac{\partial \mathbf{j}_t}{\partial t} \qquad (\text{div } \mathbf{j}_t = 0). \qquad (10.11)$$

Let us seek the solution of this equation in the form of a Fourier integral. From (10.11) we find for the Fourier components of the scattered wave field:

$$\mathbf{E}_{\mathbf{k}\omega} = i \, \frac{4\pi\omega}{c^2} \, \frac{\mathbf{j}_{\mathbf{k}\omega}}{k^2 - \epsilon(\omega)(\omega^2/c^2)} \,, \tag{10.12}$$

where the Fourier component of the current $\mathbf{j}_{\mathbf{k}\omega}$ is connected to the Fourier component of the electron-density fluctuations $\delta n_{\mathbf{q}\Delta\omega}$ by means of the relation

$$\mathbf{j}_{\mathbf{k}\omega} = i \, \frac{e^2}{m\omega_0} \, \delta n_{\mathbf{q}\Delta\omega} \mathbf{E}_0{}^\perp, \qquad \mathbf{q} = \mathbf{k} - \mathbf{k}_0, \qquad \Delta\omega = \omega - \omega_0 \tag{10.13}$$

($\mathbf{E}_0{}^\perp$ is the component of $\mathbf{E}_0$ perpendicular to the vector $\mathbf{k}$). The increment in energy of the electromagnetic field of scattered waves per unit time is determined by the expression

$$I = -\tfrac{1}{2} \, \mathrm{Re} \int d\mathbf{r} \, \mathbf{E}(\mathbf{r}, t) \, \mathbf{j}_t{}^*(\mathbf{r}, t). \tag{10.14}$$

Substituting $\mathbf{E}(\mathbf{r}, t)$ and $\mathbf{j}_t(\mathbf{r}, t)$ in the form of Fourier integrals, and averaging, we obtain

$$I = \frac{V}{(2\pi)^3} \, \frac{e^4}{m^2 c^2 \omega_0{}^2} \, \mathrm{Im} \int d\mathbf{k} \, d\omega\omega \, \frac{\langle \delta n^2 \rangle_{\mathbf{q}\Delta\omega}}{k^2 - \epsilon(\omega)(\omega^2/c^2)} \, \mathbf{E}_0{}^{\perp\,2}, \tag{10.15}$$

where V is the total volume. [In deriving (10.15) we used the relationship

$$\langle \delta n_{\mathbf{q}\Delta\omega}^*, \delta n_{\mathbf{q}\Delta\omega} \rangle = 2\pi V \delta(\omega - \omega') \langle \delta n^2 \rangle_{\mathbf{q}\Delta\omega} \, . \,]$$

Since the integrand in (10.15) is real, the contribution to the imaginary part of (10.15) is given only by the poles of the integrand. Making the formal substitution

$$\left\{ k^2 - \epsilon(\omega) \, \frac{\omega^2}{c^2} \right\}^{-1} \rightarrow i\pi\delta \left\{ k^2 - \epsilon(\omega) \, \frac{\omega^2}{c^2} \right\}$$

and integrating with respect to the modulus of the vector $\mathbf{k}$, we obtain

$$d\bar{I} = \frac{cV}{16\pi^2} \left(\frac{e^2}{mc^2} \right)^2 \frac{\omega^2}{\omega_0{}^2} \, \sqrt{\epsilon(\omega)} \, \mathbf{E}_0{}^{\perp\,2} \langle \delta n^2 \rangle_{\mathbf{q}\Delta\omega} \, d\omega \, do. \tag{10.16}$$

The vector $\mathbf{k}$ and ω in (10.16) are connected by means of the relationship $k^2 = \epsilon(\omega)(\omega^2/c^2)$, and hence $\mathbf{k}$ and ω may be considered as the wave vector and frequency of the scattered wave. Correspondingly, the quantities $\mathbf{q}$ and $\Delta\omega$ acquire the meaning of a change in the wave vector and a change in the frequency during scattering of waves. Formula (10.16) determines the wave-scattering intensity in the frequency band $d\omega$ in the solid angle element do.

Although we took into account only electromagnetic wave scattering by fluctuations in the electron density in deriving (10.16), the scattering intensity (10.16) turns out to be dependent on the ion motion also. This is because the spectral distribution of the electron-density fluctuations $\langle \delta n^2 \rangle_{\mathbf{q}\Delta\omega}$ depends in an essential manner on the ion motion because of the self-consistent interaction between electrons and ions.

Let ϑ denote the angle between the vectors $\mathbf{k}$ and $\mathbf{k}_0$, called the scattering angle. Let us select a coordinate system such that the vector $\mathbf{k}$ would be directed along the z axis, but the vector $\mathbf{k}_0$ would lie in the yz plane. Then, as is not difficult to see

$$\mathbf{E}_0^{\perp^2} = E_0^2 \{ \cos^2 \varphi + \cos^2 \vartheta \sin^2 \varphi \}, \tag{10.17}$$

where φ is the angle between the x axis and the vector $\mathbf{E}_0$. If the incident wave is not polarized, then (10.17) should be averaged over the various orientations of the vector $\mathbf{E}_0$, i.e., over the angle φ. We obtain as a result of the averaging

$$\overline{\mathbf{E}_0^{\perp^2}} = \tfrac{1}{2}(1 + \cos^2 \vartheta) E_0^2. \tag{10.18}$$

The energy flux density of the incident wave is characterized by the Poynting vector

$$S_0 = \frac{c}{8\pi} [\epsilon(\omega_0)]^{1/2} E_0^2. \tag{10.19}$$

Dividing the scattering intensity (10.16) by the energy flux density of the incident wave (10.19) and the magnitude of the scattering volume V, we find the differential scattering cross section (scattering coefficient). For an unpolarized wave the differential scattering cross section referred to the solid angle element do and the frequency band $d\omega$, equals (2)

$$d\Sigma = \frac{1}{4\pi} \left(\frac{e^2}{mc^2}\right)^2 \frac{\omega^2}{\omega_0{}^2} \left[\frac{\epsilon(\omega)}{\epsilon(\omega_0)}\right]^{1/2} (1 + \cos^2 \vartheta)\langle\delta n^2\rangle_{\mathbf{q}\Delta\omega} \, d\omega \, do, \qquad (10.20)$$

where $\epsilon(\omega) = 1 - (\Omega^2/\omega^2)$, ω_0 and $\omega > \Omega$, $\Delta\omega = \omega - \omega_0$, $\mathbf{q} = \mathbf{k} - \mathbf{k_0}$, and ϑ is the scatter angle.

Let us note that (10.20) is valid for an arbitrary frequency change during scattering. If $\Delta\omega \ll \omega_0$, the factor $(\omega^2/\omega_0{}^2)[\epsilon(\omega)/\epsilon(\omega_0)]^{1/2}$ becomes unity, and (10.20) goes over into a known formula determining the scattering by density fluctuations with a small change in frequency.

4. Total Scattering Coefficient for an Isothermal Plasma

According to (10.20), the spectral distribution of the scattered waves is determined by the spectral distribution of the electron density fluctuations in the plasma. In the case of equal electron and ion temperatures (isothermal plasma), the spectrum of scattered radiation will consist of a Doppler broadened fundamental line ($\Delta\omega \lesssim qs_i$) and sharp maximums at $\Delta\omega = \pm\Omega$ (if $aq \ll 1$). Hence, in the most interesting case of high incident frequencies ($\omega_0 \gg \Omega$), the factor $(\omega^2/\omega_0{}^2)[\epsilon(\omega)/\epsilon(\omega_0)]^{1/2}$ may be considered to equal unity, whereupon the cross section (10.20) may be integrated with respect to the frequency by using (5.29). Therefore, we obtain the following expression for the integral scattering coefficient (in the given range of angles) of an isothermal plasma (2, 4):

$$d\Sigma = \frac{1}{2} n_0 \left(\frac{e^2}{mc^2}\right)^2 \frac{1 + a^2q^2}{2 + a^2q^2} (1 + \cos^2 \vartheta) \, do, \quad q = 2\frac{\omega_0}{c}\sin\frac{\vartheta}{2}. \quad (10.21)$$

Integrating (10.21) with respect to the angles, we find the total scattering coefficient

$$\Sigma = \left\{1 - \frac{3}{4a^2k_0{}^2} + \frac{3}{8a^2k_0{}^2}\ln(1 + 2a^2k_0{}^2) \right.$$
$$\left. + \frac{3}{4}\frac{1 - a^2k_0{}^2}{\sqrt{2}a^3k_0{}^3}\arctan(\sqrt{2}ak_0)\right\} n_0\sigma_T. \qquad (10.22)$$

In the limiting cases of short and long wavelengths, we obtain the respective approximate expressions

$$\Sigma = n_0 \sigma_T, \qquad a^2 k_0^2 \gg 1, \qquad (10.23)$$

$$\Sigma = \tfrac{1}{2} n_0 \sigma_T, \qquad a^2 k_0^2 \ll 1, \qquad (10.24)$$

where $\sigma_T = (8\pi/3)(e^2/mc^2)^2$ is the Thomson scattering cross section of electromagnetic waves by a free electron.

5. Spectral Distribution of Scattering

Let us consider the spectral distribution of the scattered radiation in more detail. Let us hence take account of the possibility of a difference in the electron and ion temperatures in the plasma T_e and T_i. Substituting (5.42) into (10.20), we obtain the following general formula for the differential electromagnetic wave scattering cross section in a nonisothermal plasma:

$$\frac{d\Sigma}{dz\, do} = \frac{1}{2\sqrt{\pi}}\, n_0 \left(\frac{e^2}{mc^2}\right)^2 \frac{\omega^2}{\omega_0^2} \left(\frac{\epsilon}{\epsilon_0}\right)^{1/2} (1 + \cos^2 \vartheta)\, \Phi(z, q), \qquad (10.25)$$

$$\Phi(z, q) = \frac{\{[a^2 q^2 + t(1 - \varphi(\mu z))]^2 + \pi t^2 \mu^2 z^2 e^{-2\mu^2 z^2}\}\, e^{-z^2} + \mu\{[1 - \varphi(z)]^2 + \pi z^2 e^{-2z^2}\}\, e^{-\mu^2 z^2}}{\{a^2 q^2 + 1 - \varphi(z) + t[1 - \varphi(\mu z)]\}^2 + \pi z^2 \{e^{-z^2} + t\mu e^{-\mu^2 z^2}\}^2}, \qquad (10.26)$$

where $z = \sqrt{\tfrac{3}{2}}\,(\Delta\omega/qs)$, $t = T_e/T_i$, and $\mu^2 = t(M/m)$.

The spectral distribution of the scattered radiation $\Phi(z, q)$ is presented in Fig. 20 as a function of the nondimensional frequency z for an isothermal plasma for values of the parameter $a^2 q^2 = 0.1$; 0.5; 1; and 10. The spectral distribution of the scattered radiation $\Phi(z, q)$ is presented in Fig. 21 for a nonisothermal plasma, as a function of the nondimensional frequency z for values of the temperature ratio $t = 1; 2; 5;$ and 10 $(a^2 q^2 = 0.1)$.

Let us consider the character of the scattering in the various frequency bands in more detail. In the limiting case of short wavelengths $a^2 q^2 \gg 1$, the function (10.26) reduces to a Gaussian

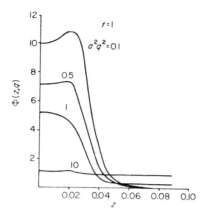

FIG. 20.

function $\Phi(z, q) \simeq e^{-z^2}$, hence the spectral distribution of the scattering is Gaussian in nature

$$d\Sigma = \sqrt{\frac{3}{8\pi}}\, n_0 \left(\frac{e^2}{mc^2}\right)^2 (ks)^{-1}(1 + \cos^2 \vartheta) \exp\left[-\frac{3}{2}\frac{\omega^2}{k^2 s^2}\right] d\omega \, do, \quad a^2 q^2 \gg 1.$$

$$(10.27)$$

The Doppler line broadening is determined by the thermal electron velocity. The total scattering cross section equals the sum of the scattering cross sections by the individual electrons (10.23). In this case the Coulomb interaction between electrons and ions is insignificant, and the scattering is incoherent.

Of more interest is the limiting case of long wavelengths $a^2 q^2 \ll 1$, when the collective properties of the plasma are manifest. (This case is realized under conditions of experiments on radiowave scattering by density fluctuations in the upper layers of the ionosphere: $\lambda \sim 10$ cm, $a \sim 1$ cm.) Utilizing the asymptotic representations (3.27) and (3.28) of the function $\varphi(z)$, formula (10.25) giving the spectral distribution of the scattered radiation, may in this case be simplified substantially for various frequency ranges.

(1) $z \ll \mu^{-1} \ll 1$ $(\varDelta\omega \ll qs_i)$. Using the expansion (3.27) for the electron and ion functions in (10.26), we have

$$\Phi(z, q) \simeq \frac{t^2 + \mu}{(1 + t)^2}.$$

In the case of a strongly nonisothermal plasma

$$\Phi(z, q) \simeq 1 + \sqrt{\frac{M}{m}}\, t^{-2/3}.$$

The scattering coefficient with a small frequency change $(\varDelta\omega \ll qs_i)$ is determined by the formula

$$d\Sigma = \frac{1}{\sqrt{8\pi}}\, n_0 \left(\frac{e^2}{mc^2}\right)^2 \frac{T_e^{3/2}\sqrt{m} + T_i^{3/2}\sqrt{M}}{q(T_e + T_i)^2}(1 + \cos^2 \vartheta)\, d\omega\, do.$$

$$(10.28)$$

Let us note that the scattering coefficient with a small frequency change is $\frac{1}{4}\sqrt{M/m}$ times less in the case of a strongly nonisothermal plasma than the corresponding coefficient for an isothermal plasma.

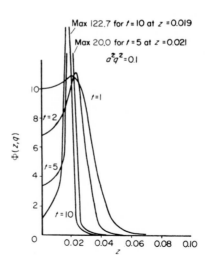

Fig. 21.

(2) $z \lesssim \mu^{-1} \ll 1$ $(\Delta\omega \lesssim qs_i \ll qs)$. Having utilized the expansion (3.28) only for the electron function in (10.26), we have

$$\Phi(z, q) \simeq \frac{\{a^2q^2 + t[1 - \varphi(\mu z)]\}^2 + \pi t^2\mu^2 z^2 e^{-2\mu^2 z^2} + \mu e^{-\mu^2 z^2}}{\{a^2q^2 + 1 + t[1 - \varphi(\mu z)]\}^2 + \pi t^2\mu^2 z^2 e^{-2\mu^2 z^2}}.$$

If $\mu \gg t^2$ $(\sqrt{M/m} \gg t^{3/2})$, then

$$\Phi(z, q) \simeq \frac{\mu \exp[- \mu^2 z^2]}{\{1 + t[1 - \varphi(\mu z)]\}^2 + \pi t^2\mu^2 z^2 \exp[- 2\mu^2 z^2]}.$$

In particular, in the isothermal case $(t = 1)$

$$\Phi(z, q) \simeq \frac{\mu \exp[- \mu^2 z^2]}{[2 - \varphi(\mu z)]^2 + \pi\mu^2 z^2 \exp[- 2\mu^2 z^2]}. \tag{10.29}$$

The function (10.29) is characterized by a steep drop at $z \simeq 1/\mu$, i.e., for a $\Delta\omega \simeq qs_i$ change in frequency. This drop indeed determines the Doppler width of the spectral distribution of the scattered radiation in the isothermal plasma. The width is therefore determined by the thermal ion velocity despite the fact that the scattering occurs on the electrons. Substituting (10.29) into (10.25) we obtain the following formula for the scattering coefficient

$$d\Sigma = \sqrt{\frac{3}{8\pi}}\, n_0 \left(\frac{e^2}{mc^2}\right)^2 (qs_i)^{-1}$$

$$\times \frac{\exp\left[-\dfrac{3}{2}\dfrac{\Delta\omega^2}{q^2 s_i^2}\right]}{\left[2 - \varphi\left(\sqrt{\dfrac{3}{2}}\dfrac{\Delta\omega}{qs_i}\right)\right]^2 + \dfrac{3\pi\Delta\omega^2}{2q^2 s_i^2 e}\exp\left[-3\dfrac{\Delta\omega^2}{q^2 s_i^2}\right]}(1 + \cos^2 \vartheta)\, d\omega\, do. \tag{10.30}$$

Since the frequency range $\Delta\omega$ less than or on the order of qs_i yields the principal contribution to the integral cross section, (10.30) turns out to be a good approximation to the scattering coefficient in the case of an isothermal plasma in the whole frequency band.

If $t^2 \gg \mu$ (strongly nonisothermal plasma $t^{3/2} \gg \sqrt{M/m}$),

$$\Phi(z, q) \cong \frac{t^2\{[1 - \varphi(\mu z)]^2 + \pi\mu^2 z^2 \exp[- 2\mu^2 z^2]\}}{\{1 + t[1 - \varphi(\mu z)]\}^2 + \pi t^2\mu^2 z^2 \exp[- 2\mu^2 z^2]} \cong 1.$$

Therefore, in a nonisothermal plasma the scattering coefficient does not change essentially in the range of frequency changes $\Delta\omega \lesssim qs_i$.

(3) $\mu^{-1} \ll z \ll 1$ $(qs_i \ll \Delta\omega \ll qs)$. Utilizing the expansion (3.27) for the electrons and (3.28) for the ions, we have

$$\Phi(z, q) \simeq \frac{\left(a^2q^2 - \dfrac{t}{2\mu^2 z^2}\right)^2 + \mu \exp[-\mu^2 z^2]}{\left(a^2q^2 + 1 - \dfrac{t}{2\mu^2 z^2}\right)^2 + \pi z^2}.$$

If $\mu \gg t^2 \sim 1$, then $\Phi(z, q) \simeq a^4 q^4$ and the scattering cross section is very small. However, in the case of a strongly nonisothermal plasma $(t^2 \gg \mu)$ the ratio $t/2\mu^2 z^2$ may take on a value on the order of unity, and the function $\Phi(z, q)$ has a sharp maximum at $z^2 = t/2\mu^2$:

$$\Phi(z, q) \cong \frac{t^2/4\mu^4}{[z^2 - (t/2\mu^2)]^2 + \pi z^6} \simeq \sqrt{\pi} z \delta\left(z^2 - \frac{t}{2\mu^2}\right), \quad z \ll 1. \quad (10.31)$$

The condition $z^2 = t/2\mu^2$ means that $\Delta\omega = \pm\omega_s(q)$, where $\omega_s(q) = qs$ is the frequency of the sonic oscillations being propagated in the strongly nonisothermal plasma. Substituting (10.31) into (10.25), we have

$$d\Sigma = \frac{1}{4} n_0 \left(\frac{e^2}{mc^2}\right)^2 (1 + \cos^2\vartheta)\{\delta(\Delta\omega - qs) + \delta(\Delta\omega + qs)\}\, d\omega\, do,$$

$$t^{3/2} \gg \sqrt{\frac{M}{m}}. \quad (10.32)$$

Thus, the coefficient of electromagnetic wave scattering in a nonisothermal plasma has sharp maximums at $\Delta\omega = \pm qs$, associated with the excitation of sonic oscillations in the plasma.

Integrating (10.32) with respect to the angles and frequencies, we find the total scattering coefficient in a nonisothermal plasma

$$\Sigma = n\sigma_T, \quad t^{3/2} \gg \sqrt{\frac{M}{m}}. \quad (10.33)$$

This coefficient turns out to be twice the coefficient (10.24) in the isothermal case.

(4) $\mu^{-1} \ll z \lesssim 1$ $(qs_i \ll \Delta\omega \lesssim qs)$. Utilizing the expansion (3.28) for the ions we have

$$\Phi(z, q) \simeq \frac{[a^2q^2 - (t/2\mu^2 z^2)]^2 \, e^{-z^2}}{[a^2q^2 + 1 - \varphi(z) - (t/2\mu^2 z^2)]^2 + \pi z^2 e^{-z^2}} \simeq O(a^4 q^4).$$

The scattering cross section in this frequency range is very small in both the isothermal and nonisothermal cases.

(5) $z \gg 1$ $(\Delta\omega \gg qs)$. Using the expansion (3.28) for the electrons and ions, and remarking that $\mu \gg 1$, we have

$$\Phi(z, q) \simeq \frac{a^4 q^4 e^{-z^2}}{[a^2q^2 - (1/2z^2)]^2 + \pi z^2 e^{-2z^2}} \simeq \sqrt{\pi} \, a^2 q^2 z \delta \left(z^2 - \frac{1}{2a^2 q^2}\right).$$

The function $\Phi(z, q)$ has a sharp maximum for $z^2 = 1/2a^2 q^2$, i.e., at the frequencies $\Delta\omega = \pm\Omega$. Hence, if $\Delta\omega \gg qs$, electromagnetic wave scattering occurs at the plasma oscillations. For an arbitrary relationship between the electron and ion temperatures T_e and T_i, the differential scattering cross section in this frequency range has the form

$$d\Sigma = \frac{1}{4} n_0 \left(\frac{e^2}{mc^2}\right)^2 \frac{\omega^2}{\omega_0^2} \sqrt{\frac{\epsilon}{\epsilon_0}} \, a^2 q^2 (1 + \cos^2 \vartheta) \{\delta(\Delta\omega - \Omega)$$
$$+ \, \delta(\Delta\omega + \Omega)\} \, d\omega \, do. \tag{10.34}$$

Correspondingly, the integral scattering cross section of electromagnetic waves by plasma oscillations in the high-frequency case $(\omega_0 \gg \Omega)$ is

$$d\Sigma = \frac{1}{2} n_0 \left(\frac{e^2}{mc^2}\right)^2 a^2 q^2 (1 + \cos^2 \vartheta) \, do, \quad q = 2k_0 \sin \frac{\vartheta}{2} ;$$
$$\Sigma = 2a^2 k_0^2 n_0 \sigma_T . \tag{10.35}$$

An analysis of the case $a^2 k_0^2 \ll 1$ shows that the electromagnetic wave scattering spectra differ essentially for isothermal and nonisothermal conditions in the plasma. In an isothermal plasma there is a central maximum in the scattering spectrum, due to incoherent scattering by electron-density fluctuations with a Doppler width determined by the ion velocities, and side satellites due to scattering by the plasma oscillations. The relative weight of the satellites relative to the principal maximum is $4a^2k_0{}^2$. In a strongly nonisothermal plasma there is no central maximum. There are two maximums, symmetrically located with respect to $\varDelta\omega = 0$ which are due to scattering by the sonic oscillations, and side satellites associated with the scattering by the plasma oscillations. The weight of the plasma satellites relative to the sonic maximums is $2a^2k_0{}^2$.

The spectral distribution for the intermediate cases is pictured in Fig. 21. Let us note that an experimental study of the shape of the electromagnetic wave scattering spectrum in a plasma may be an important source of information on the parameters characterizing the state of the plasma. In particular, such a study might permit a conclusion to be drawn not only on the electron density and temperature, but also on the degree of nonisothermy of the plasma.

6. Transformation of Transverse into Longitudinal Electromagnetic Waves

Let us determine the electromagnetic wave absorption associated with the excitation of plasma oscillations, for which the energy of the incident transverse electromagnetic wave goes over into the energy of the longitudinal plasma oscillations. Keeping only the longitudinal part of the current $\mathbf{j}_l$ (rot $\mathbf{j}_l = 0$) in (10.10), we obtain the following expressions directly from (10.10) for the Fourier component of the longitudinal component of the transformed wave field:

$$E_{\mathbf{k}\omega}^l = -\frac{4\pi i}{\omega\epsilon_l(\omega,\mathbf{k})}j_{\mathbf{k}\omega}^l, \quad j_{\mathbf{k}\omega}^l = i\frac{e^2}{m\omega_0}\left(1 + \frac{\varDelta\omega}{\omega}\frac{k^2}{q^2}\right)\delta n_{\mathbf{q}\varDelta\omega}E_0{}^l, \quad (10.36)$$

where $E_0{}^l$ is the component of the incident wave field along the vector $\mathbf{k}$. Let us note that it is necessary to take account of spatial dispersion in (10.36), i.e., the dependence of the plasma dielectric permittivity ϵ_l on the wave vector $\mathbf{k}$, since the group velocity of the longitudinal waves turns out to be zero if such a dependence is neglected.

Determining the energy increment of the field of scattered longitudinal waves exactly as for the transverse case, we obtain the following expression for the excitation intensity after averaging:

$$\langle I \rangle = \frac{V}{(2\pi)^3} \frac{e^4}{m^2 \omega_0{}^2} \operatorname{Re} i \int d\mathbf{k} \, d\omega \frac{\langle \delta n^2 \rangle_{\mathbf{q} \Delta \omega}}{\omega \epsilon_l(\omega, \mathbf{k})} \left(1 + \frac{\Delta\omega}{\omega} \frac{k^2}{q^2} \right)^2 E_0^{l^2}. \qquad (10.37)$$

Only frequencies for which the denominator of the integrand vanishes $(\epsilon_l(\omega, \mathbf{k}) = 0)$, yield a contribution to (10.37), i.e., longitudinal plasma waves are excited as a result of absorption by the transverse electromagnetic wave. The dispersion law for plasma waves may be written in the approximation (3.33) as

$$k^2 = \frac{\omega^2}{s^2} \epsilon(\omega), \qquad (10.38)$$

where $\epsilon(\omega) = 1 - (\Omega^2/\omega^2)$. Hence, the intensity of plasma-wave excitation equals

$$\langle dI \rangle = \frac{V}{16\pi^2} \frac{e^4}{m^2 \omega_0{}^2} \frac{\omega^2}{s^3} \sqrt{\epsilon(\omega)} \left(1 + \frac{\Delta\omega}{\omega} \frac{k^2}{q^2} \right)^2 E_0^{l^2} \langle \delta n^2 \rangle_{\mathbf{q} \Delta \omega} \, d\omega \, do. \qquad (10.39)$$

If the incident wave is unpolarized, then

$$E_0^{l^2} = \tfrac{1}{2} E_0{}^2 \sin^2 \vartheta.$$

Dividing (10.39) by the energy flux density of the incident wave (10.19) and the volume V, we obtain the differential coefficient of electromagnetic wave transformation into plasma waves as

$$d\Sigma_{E \to P} = \frac{1}{4\pi} \left(\frac{e^2}{mc^2} \right)^2 \frac{c^3}{s^3} \frac{\omega^2}{\omega_0{}^2} \sqrt{\frac{\epsilon(\omega)}{\epsilon(\omega_0)}} \left(1 + \frac{\Delta\omega}{\omega} \frac{k^2}{q^2} \right)^2 \sin^2 \vartheta \langle \delta n^2 \rangle_{\mathbf{q} \Delta \omega} \, d\omega \, do.$$

$$(10.40)$$

Formula (10.40) is applicable only in the range of frequencies ω near the Langmuir frequency Ω for which the damping of the plasma waves is slight. The ratio of the transformation coefficient (10.40) to the scattering coefficient (10.20) equals

$$\frac{d\Sigma_{E \to P}}{d\Sigma} = \frac{c^3}{s^3} \frac{\sin^2 \vartheta}{1 + \cos^2 \vartheta} \left(1 + \frac{\Delta\omega}{\omega} \frac{k^2}{q^2} \right)^2. \qquad (10.41)$$

Therefore, in the range of frequencies ω near Ω, the absorption associated with transformation of transverse into longitudinal waves is more important, as compared with the scattering of the transverse waves.

7. Longitudinal Wave Transformation and Scattering in a Plasma

In conclusion, let us consider the transformation and scattering of longitudinal plasma waves by density fluctuations in the plasma. Let us examine a plane longitudinal wave being propagated with frequency ω_0 and wave vector $\mathbf{k}_0$ in a plasma. The group velocity for the longitudinal waves in a plasma equals, according to (3.33)

$$\frac{d\omega_0}{dk_0} = s[\epsilon(\omega_0)]^{1/2}.$$

This velocity directly determines the energy flux associated with the wave. The energy flux density equals

$$S_0 = \frac{s}{8\pi} [\epsilon(\omega_0)]^{1/2} E_0^2. \tag{10.42}$$

The scattered wave field is determined by the current

$$\mathbf{j}_{\mathbf{k}\omega} = i \frac{e^2}{m\omega_0} \delta n_{\mathbf{q}\Delta\omega} \left\{ \mathbf{E}_0 + \frac{\Delta\omega}{\omega} \frac{1}{q^2} (\mathbf{q}\mathbf{E}_0) \mathbf{k} + \frac{\Delta\omega}{\omega_0} \frac{1}{q^2} (\mathbf{k}_0\mathbf{E}_0)\mathbf{q} \right\}, \tag{10.43}$$

where $\mathbf{E}_0$ is the incident wave field $(\mathbf{E}_0 \| \mathbf{k}_0)$. Taking account of the transverse part of the current (relative to $\mathbf{k}$) it is possible to obtain the following formula for the coefficient of transformation of longitudinal into transverse waves by fluctuations in a plasma:

$$d\Sigma_{P \to E} = \frac{1}{2\pi} \left(\frac{e^2}{mc^2}\right)^2 \frac{c}{s} \frac{\omega^2}{\omega_0^2} \left[\frac{\epsilon(\omega)}{\epsilon(\omega_0)}\right]^{1/2}$$

$$\times \left(1 - \frac{\Delta\omega}{\omega_0} \frac{k_0^2}{q^2}\right)^2 \sin^2 \vartheta \langle \delta n^2 \rangle_{\mathbf{q}\Delta\omega} \, d\omega \, do. \tag{10.44}$$

Since the spectral distribution of the density fluctuations is characterized by maximums at $\Delta\omega = 0$ and $\Delta\omega = \Omega$, the transverse electromagnetic waves will be radiated mainly with frequencies Ω and 2Ω.

The longitudinal part of the current (10.43) characterizes longitudinal wave scattering by density fluctuations in the plasma. The scattering coefficient of plasma waves by density fluctuations in a plasma equals

$$
d\Sigma_{P\to P} = \frac{1}{2\pi} \left(\frac{e^2}{mc^2}\right)^2 \frac{c^4}{s^4} \frac{\omega^2}{\omega_0^2} \left[\frac{\epsilon(\omega)}{\epsilon(\omega_0)}\right]^{1/2} \left(\cos^2 \vartheta + \frac{\Delta\omega}{\omega} \frac{k^2 \cos \vartheta - k_0 k}{q^2}\right.
$$
$$
\left. + \frac{\Delta\omega}{\omega_0} \frac{k_0 k - k_0^2 \cos \vartheta}{q^2}\right)^2 \langle \delta n^2 \rangle_{\mathbf{q}\Delta\omega}\, d\omega\, do. \tag{10.45}
$$

The ratio of the scattering coefficient of longitudinal waves to the transformation coefficient with a small change in frequency equals $(c^3/s^3) \cot^2 \vartheta$.

REFERENCES

1. A. I. Akhiezer, I. G. Prokhoda, and A. G. Sitenko, *ZETF* **33**, 750 (1957) *(Sov. Phys. JETP)*.
2. A. I. Akhiezer, I. A. Akhiezer, and A. G. Sitenko, *ZETF* **41**, 644 (1961) *(Sov. Phys. JETP)*.
3. F. G. Bass and A. Ia. Blank, *ZETF* **43**, 1479 (1962) *(Sov. Phys. JETP)*.
4. J. P. Dougherty and D. T. Farley, *Proc. Roy. Soc. (London)* **A259**, 79 (1960).

Scattering of Electromagnetic Waves by Fluctuations in a Plasma in the Presence of a Magnetic Field

1. General Formula for the Scattering Cross Section in the Presence of a Magnetic Field

Now let us consider the influence of an external magnetic field on the electromagnetic wave scattering in a plasma (1, 2). The presence of the external magnetic field in the plasma leads to splitting of the frequency of the plasma oscillations, and also to the appearance of new kinds of proper oscillations in the low-frequency range of the spectrum (Alfven and magnetosonic waves). Hence, additional maximums associated with the scattering by the Alfven and magnetosonic oscillations, originate in the spectrum of scattered radiation in a plasma in the presence of a magnetic field. Let us note that the electromagnetic wave scattering in the presence of a magnetic field is due not only to electron density fluctuations, but also to electron velocity fluctuations and magnetic field fluctuations.

Neglecting ion motion the electromagnetic field in a plasma is determined by the system of equations:

$$\text{rot } \mathbf{E} = -\frac{1}{c}\frac{\partial B}{\partial t}, \qquad \text{rot } \mathbf{B} = \frac{1}{c}\frac{\partial \mathbf{E}}{\partial t} + \frac{4\pi}{c}\, en\mathbf{v},$$

$$\frac{d\mathbf{v}}{dt} + (\mathbf{v}\nabla)\mathbf{v} = \frac{e}{m}\mathbf{E} + \frac{e}{mc}\,[\mathbf{v}, \mathbf{B}], \qquad \frac{\partial n}{\partial t} + \text{div } n\mathbf{v} = 0, \tag{11.1}$$

where $\mathbf{E}$ and $\mathbf{B}$ are the electric and magnetic fields ($\mathbf{B}$ also includes the external permanent magnetic field $\mathbf{B}_0$), $\mathbf{v}$ and n are the velocity and density of the plasma electrons. If the electron density is constant n_0, then by linearizing the last of equations (11.1), the system may be reduced to the wave equation

$$\left\{ \delta_{ij}\Delta - \nabla_i\nabla_j - \frac{\epsilon_{ij}}{c^2}\frac{\partial^2}{\partial t^2} \right\} E_j = 0, \tag{11.2}$$

where ϵ_{ij} is the dielectric permittivity tensor of a plasma in a magnetic field (6.22). Equation (11.2) admits of a solution in the form of plane monochromatic waves with a dispersion law and polarization determined by (6.25) and (6.26).

Let a plane monochromatic wave of definite kind (ordinary or extraordinary) be propagated in a plasma

$$\mathbf{E}_0(\mathbf{r}, t) = e_0 E_0 e^{i\mathbf{k}_0\mathbf{r} - i\omega_0 t}, \tag{11.3}$$

where E_0 is the real amplitude of the wave. Because of fluctuations in the electron density, the electron velocity, and the magnetic field δn, $\delta \mathbf{v}$, and $\delta \mathbf{B}$ during propagation of this wave in the plasma, scattered waves originate whose field will be determined, according to (11.1), by the equation:

$$\left\{ \delta_{ij}\Delta - \nabla_i\nabla_j - \frac{\epsilon_{ij}}{c^2}\frac{\partial^2}{\partial t^2} \right\} E_j = \frac{4\pi}{c^2}\frac{\partial j_i}{\partial t}, \tag{11.4}$$

where $\mathbf{j}(\mathbf{r}, t)$ is the current due to the incident wave field and the thermal fluctuations in the plasma. The space-time Fourier component of the current $\mathbf{j}(\mathbf{k}, \omega)$ is connected to the corresponding components $\delta n(\mathbf{q}, \Delta\omega)$, $\delta \mathbf{v}(\mathbf{q}, \Delta\omega)$, and $\delta \mathbf{B}(\mathbf{q}, \Delta\omega)$ by means of the relationship:

$$\begin{aligned}
j_i(\mathbf{k}, \omega) = &-\frac{i\omega_0}{4\pi}\left\{ (\epsilon_{ij}^0 - \delta_{ij})\frac{\delta n(\mathbf{q}, \Delta\omega)}{n_0} - i\frac{e}{mc}\frac{\omega}{\Omega^2}(\epsilon_{il} - \delta_{il}) \right.\\
&\times (\epsilon_{kj}^0 - \delta_{kj})\,\epsilon_{lkm}\delta B_m(\mathbf{q}, \Delta\omega)\\
&+ \frac{1}{\omega_0}\left[\frac{\omega}{\omega_0}(\epsilon_{ik} - \delta_{ik})(k_k^0\delta_{jl} - \delta_{kj}k_l^0) \right.\\
&- \frac{\omega_0^2}{\Omega^2}(\epsilon_{mj}^0 - \delta_{mj})(q_m\delta_{kl} + \delta_{mk}k_l^0)\\
&\left.\left. + (\epsilon_{kj}^0 - \delta_{kj})\,k_k^0\delta_{il} \right] \delta v_l(\mathbf{q}, \Delta\omega) \right\} E_j^0,
\end{aligned} \tag{11.5}$$

where $\mathbf{q} = \mathbf{k} - \mathbf{k}_0$, $\Delta\omega = \omega - \omega_0$, $\epsilon_{ij}^0 = \epsilon_{ij}(\omega_0)$, and $\epsilon_{ij} = \epsilon_{ij}(\omega)$. We shall seek the solution of (11.4) in the form

$$\mathbf{E}(\mathbf{r}, t) = \sum_{\lambda=\pm} \int \frac{d\mathbf{k}_\lambda}{(2\pi)^3} \, \mathbf{e}^\lambda q_{\mathbf{k}_\lambda}(t) e^{i\mathbf{k}_\lambda \mathbf{r}}. \tag{11.6}$$

Substituting (11.6) into (11.4) and using the condition of orthogonality of the polarizations for the ordinary and extraordinary waves of a given frequency, we obtain the following expression for the Fourier components of the coefficient $q_{\mathbf{k}_\lambda}(t)$ in the expansion (11.6):

$$q_{\mathbf{k}_\lambda}(\omega) = \frac{4\pi i \omega}{\epsilon_{ij} e_i^{\lambda*} e_j^\lambda} \cdot \frac{j_l(\mathbf{k}_\lambda, \omega) \, e_l^{\lambda*}}{(k_\lambda^2 c^2 / \eta_\lambda^2) - \omega^2}. \tag{11.7}$$

Hence, the following relationship

$$E_i(\mathbf{k}, \omega) = \sum_{\lambda=\pm} \frac{4\pi i}{\epsilon_{ij} e_i^{\lambda*} e_j^\lambda} \frac{\omega}{(k^2 c^2 / \eta_\lambda^2) - \omega^2} j_l(\mathbf{k}, \omega) \, e_l^{\lambda*} e_i^\lambda \tag{11.8}$$

holds between the scattered wave field $\mathbf{E}(\mathbf{k}, \omega)$ and the forcing current $\mathbf{j}(\mathbf{k}, \omega)$.

The energy increment of the scattered wave field per unit time is given by (10.14). Substituting $\mathbf{j}(\mathbf{r}, t)$ and $\mathbf{E}(\mathbf{r}, t)$ in the form of Fourier integrals into (10.14) and taking the statistical average, we find

$$I = \frac{V}{(2\pi)^3} \sum_{\lambda=\pm} \mathrm{Im} \int d\mathbf{k}\, d\omega \, \frac{\omega}{\epsilon_{ij} e_i^{\lambda*} e_j^\lambda} \frac{\langle |\, \mathbf{j} \mathbf{e}^{\lambda*} \,|^2 \rangle_{\mathbf{k}\omega}}{(k^2 c^2 / \eta_\lambda^2) - \omega^2}, \tag{11.9}$$

where V is the total volume. Since the integrand in (11.9) is real, only the poles of the integrand yield a contribution after integration with respect to the modulus of the vector $\mathbf{k}$. As a result, we obtain the following formula for the scattering intensity of waves of a definite kind λ:

$$dI^\lambda = \frac{V}{16\pi^2} \frac{\eta_\lambda^3}{c^3} \frac{\omega^2}{\epsilon_{ij} e_i^{\lambda*} e_j^\lambda} \langle |\, \mathbf{j} \mathbf{e}^{\lambda*} \,|^2 \rangle_{\mathbf{k}\omega} \, d\omega \, do, \quad \lambda = \pm. \tag{11.10}$$

The vector $\mathbf{k}$ and ω in (11.10) are connected by means of the relationship $k^2 = (\omega^2/c^2) \, \eta_\lambda^2$ and, hence, $\mathbf{k}$ and ω may be considered as the wave vector and frequency of the scattered wave. The energy flux density of the incident wave S_0 is determined by the projections of the Poynting vector along the vector $\mathbf{k}_0$:

$$S_0 = \frac{c}{8\pi} \eta_0 E_0^2 \left\{ |\mathbf{e}_0|^2 - \frac{|\mathbf{e}_0 \mathbf{k}_0|^2}{k_0^2} \right\},$$

$$\mathbf{e}_0 = \left(1; \frac{i\epsilon_2^0}{\eta_0^2 - \epsilon_1^0} ; \frac{\eta_0^2 \sin\vartheta_0 \cos\vartheta_0}{\eta_0^2 \sin^2\vartheta_0 - \epsilon_3^0} \right),$$

(11.11)

where ϑ_0 is the angle between $\mathbf{k}_0$ and $\mathbf{B}_0$. Dividing the scattering intensity (11.10) by the flux density (11.11) and the value of the scattering volume V, we find the differential scattering cross section (scattering coefficient). By using (11.5) we obtain the following general formula for the differential scattering cross section of electromagnetic waves in a plasma is the presence of a magnetic field (2):

$$d\Sigma = \frac{1}{2\pi} \left(\frac{e^2}{mc^2} \right)^2 \frac{\omega_0^2 \omega^2}{\Omega^4} R \left\{ |\xi|^2 \langle\delta n^2\rangle_{\mathbf{q}\Delta\omega} + \frac{n_0}{4\pi mc^2} \frac{\omega^2}{\Omega^2} a_i a_j^* \langle \delta B_i \delta B_j\rangle_{\mathbf{q}\Delta\omega} \right.$$

$$+ \frac{n_0^2}{c^2} \frac{\omega^2}{\omega_0^2} b_i b_j^* \langle \delta v_i \delta v_j\rangle_{\mathbf{q}\Delta\omega} - 2 \frac{en_0}{mc} \frac{\omega}{\Omega^2} \mathrm{Im}\, (\xi a_i^* \langle \delta n\, \delta B_i\rangle_{\mathbf{q}\Delta\omega})$$

$$+ 2 \frac{n_0}{c} \frac{\omega}{\omega_0} \mathrm{Re}\, (\xi b_i^* \langle \delta n\, \delta v_i\rangle_{\mathbf{q}\Delta\omega})$$

$$+ 2 \frac{en_0^2}{mc^2} \frac{\omega^2}{\omega_0\Omega^2} \mathrm{Im}\, (a_i b_j^* \langle\delta B_i\, \delta v_j\rangle_{\mathbf{q}\Delta\omega}) \right\} d\omega\, do,$$

(11.12)

where the following notation has been introduced:

$$R = \frac{\eta^3}{\eta_0(|\mathbf{e}_0|^2 - \{|\mathbf{e}_0 \mathbf{k}_0|^2/k_0^2\}) e_i^* \epsilon_{ij} e_j} ,$$

(11.13)

$$\xi = e_i^*(\epsilon_{ij}^0 - \delta_{ij}) e_j^0, \qquad a_i = \epsilon_{ikl} e_j^*(\epsilon_{jk} - \delta_{jk})(\epsilon_{lm}^0 - \delta_{lm}) e_m^0,$$

$$b_i = \frac{c}{\omega_0} \left\{ (\epsilon_{lk} - \delta_{lk}) \left[k_k{}^0 \delta_{ij} - \delta_{jk} k_i{}^0 - \frac{\omega_0^2}{\Omega^2} (\epsilon_{mj}^0 - \delta_{mj})(q_m \delta_{ik} + \delta_{km} k_i{}^0) \right] \right.$$

$$\left. + \frac{\omega_0}{\omega} (\epsilon_{kj}^0 - \delta_{kj}) k_k{}^0 \delta_{il} \right\} e_j^0 e_l^*,$$

$$\mathbf{e} = \left(\cos\varphi - \frac{i\epsilon_2}{\eta^2 - \epsilon_1} \sin\varphi; \sin\varphi + \frac{i\epsilon_2}{\eta^2 - \epsilon_1} \cos\varphi; \frac{\eta^2 \sin\vartheta \cos\vartheta}{\eta^2 \sin^2\vartheta - \epsilon_3} \right)$$

and ϑ is the angle between $\mathbf{k}$ and $\mathbf{B}_0$, φ the angle between the $(\mathbf{k}_0, \mathbf{B}_0)$ and $(\mathbf{k}, \mathbf{B}_0)$ planes. We omitted the superscript λ in (11.12) in order to simplify the writing.

The correlation functions of the electron-density, electron-velocity, and the magnetic field, which enter into the scattering coefficient (11.12), are expressed, according to (5.60)–(5.62), in terms of the electron and total current correlators (5.57)–(5.59). Let us note that spatial dispersion plays an essential part in the analysis of fluctuations in plasma, hence it is necessary to use the general formulas (7.3) for the dielectric permittivity, rather than (6.22), in calculating the correlators.

2. Electromagnetic Wave Scattering by Density Fluctuations

The first component in the braces, describing the wave scattering by density fluctuations,* yields a fundamental contribution to the scattering cross section (11.12). Neglecting scattering by fluctuations in the electron velocity and in the magnetic field, the scattering cross section may be written as:

$$d\Sigma = \frac{1}{2\pi} \left(\frac{e^2}{mc^2} \right)^2 \frac{\omega_0{}^2 \omega^2}{\Omega^4} R \mid \xi \mid^2 \langle \delta n^2 \rangle_{\mathbf{q} \Delta \omega} \, d\omega \, do, \qquad (11.14)$$

where the correlation function of the electron-density fluctuations is determined by the expression

$$\langle \delta n^2 \rangle_{\mathbf{q} \Delta \omega} = \frac{2}{e^2 \Delta \omega} \, \text{Im} \, \{ T_e (q_m - 4\pi q_i \chi^e_{ik} \Lambda^{-1}_{km})^* (q_n - 4\pi q_j \chi^e_{jp} \Lambda^{-1}_{pn}) \, \chi^{e*}_{mn}$$

$$+ 16\pi^2 T_i (q_i \chi^e_{ik} \Lambda^{-1}_{km})^* (q_j \chi^e_{jp} \Lambda^{-1}_{pn}) \, \chi^{i*}_{mn} \}. \qquad (11.15)$$

The factor $R \mid \xi \mid^2$ in (11.14) depends on the propagation directions of the incident and reflected waves relative to the magnetic field.

* Wave scattering by density fluctuations has also been considered in References 3 and 4.

For example, if the incident wave is propagated along the magnetic field, this factor equals

$$(R \mid \xi \mid^2)^{\pm} = \frac{\eta^3}{2\eta_0} \frac{(1 - \epsilon_1{}^0 \pm \epsilon_2{}^0)^2 \left(1 \pm \dfrac{\epsilon^2}{\eta^2 - \epsilon_1}\right)^2}{\epsilon_1 + \epsilon_1 \dfrac{\epsilon_2{}^2}{(\eta^2 - \epsilon_1)^2} + \epsilon_3 \dfrac{\eta^4 \sin^2 \vartheta \cos^2 \vartheta}{(\eta^2 \sin^2 \vartheta - \epsilon_3)^2} + \dfrac{2\epsilon_2{}^2}{\eta^2 - \epsilon_1}},$$

$$(11.16)$$

where the $(+)$ and $(-)$ signs refer, respectively, to right and left polarizations of the incident wave.

In the limiting case $q^2 c^2 / \Delta \omega^2 \gg 1$ the correlation function of the electron density fluctuations in the presence of a magnetic field (11.15) simplifies essentially, and is determined by the same expression as for a free plasma

$$\langle \delta n^2 \rangle_{\mathbf{q} \Delta \omega} = \frac{2q^2}{e^2 \Delta \omega \mid \epsilon \mid^2} \{ T_e \mid 1 + 4\pi \chi^i \mid^2 \operatorname{Im} \chi^e + 16\pi^2 T_i \mid \chi^e \mid^2 \operatorname{Im} \chi^i \},$$

$$(11.17)$$

where χ and ϵ should be understood to be the longitudinal components of the appropriate tensors, i.e., $\chi = (q_i q_j / q^2) \chi_{ij}$ and $\epsilon = (q_i q_j / q^2) \epsilon_{ij}$. Utilizing (7.3), the following expressions may be obtained for these quantities:

$$\epsilon = 1 + 4\pi \sum_{\alpha} \chi^{\alpha}, \qquad \alpha = e, i;$$

$$\chi^{\alpha} = \frac{1}{4\pi a^2 q^2} \left\{ 1 - e^{-\beta_{\alpha}} \sum_{n} \frac{z_0{}^{\alpha}}{z_n{}^{\alpha}} I_n(\beta_{\alpha})[\varphi(z_n{}^{\alpha}) - i \sqrt{\pi} z_n{}^{\alpha} \exp(- z_n{}^{\alpha 2})] \right\}.$$

$$(11.18)$$

If the change in the wave vector during scattering $\mathbf{q}$ is parallel to $\mathbf{B}_0$, the spectral distribution of the scattered radiation is exactly the same as in the absence of a magnetic field. If the direction of $\mathbf{q}$ does not agree with that of $\mathbf{B}_0$, the magnetic field influences the scattered radiation spectrum. However, this influence turns out to be slight in the domain of small frequency shifts, and resonance effects for frequency shifts which are approximately multiples of the ion cyclotron frequency originate only for angles between $\mathbf{q}$ and $\mathbf{B}_0$, which are near $\pi/2$.

In the isothermal plasma case ($T_e = T_i$) the scattered radiation spectrum for angles between $\mathbf{q}$ and $\mathbf{B}_0$, which differ from $\pi/2$ is characterized by a sharp maximum for $\Delta\omega = 0$, exactly as in the absence of a magnetic field. For $T_e = T_i$ and $\omega_0 \gg \omega_\pm$ ($\omega_\pm$ the frequencies of the Langmuir oscillations of the plasma in the magnetic field), an integral scattering coefficient may be found (in the given range of angles) by using the dispersion relation for (11.17). The integral scattering coefficient turns out to equal (1)*:

$$d\Sigma = \frac{1}{2} n_0 \left(\frac{e^2}{mc^2}\right)^2 \frac{\omega_0^4}{\Omega^4} (R \mid \xi \mid^2)_{\omega=\omega_0} \frac{1 + a^2 q^2}{2 + a^2 q^2} \, do. \qquad (11.19)$$

For $a^2 q^2 \gg 1$ formula (11.19) may be considered as an extension of the Rayleigh formula to the case of a magnetoactive medium.

The terms discarded in (11.12) yield a relative contribution on the order of s/c and s^2/c^2 to the integral scattering coefficient. Nevertheless, to neglect terms in the differential scattering coefficient which are connected with the scattering by electron-density and magnetic-field fluctuations is impossible since these processes may turn out to be significant in narrow frequency bands corresponding to the Alfven and magnetosonic oscillations.

In a nonisothermal plasma the maximum in the scattered radiation spectrum, due to the interaction with noncoherent fluctuations, is reduced radically. In a quite nonisothermal plasma ($T_e \gg T_i$) the height of the maximum is $(M/m)^{1/2}$ times less as compared with the isothermal case.

Transformation of electromagnetic into Langmuir waves is also possible because of the electromagnetic wave interaction with density fluctuations in the plasma. The differential transformation cross section of the ordinary or extraordinary waves into a Langmuir wave is

$$d\Sigma = \frac{1}{2\pi} \left(\frac{e^2}{mc^2}\right)^2 \frac{\omega_0^2 \omega^2}{\Omega^4} \frac{\eta_L^3 \mid \mathbf{k}(\hat{\epsilon}_0 - 1)\mathbf{e}_0 \mid^2}{\eta_0 \left(\mid \mathbf{e}_0 \mid^2 - \dfrac{\mid \mathbf{k}_0 \mathbf{e}_0 \mid^2}{k_0^2}\right) (\mathbf{k}\hat{\epsilon}\mathbf{k})} \langle \delta n^2 \rangle_{q\Delta\omega} \, d\omega \, do . \qquad (11.20)$$

Exactly as in the scattering case, interaction with noncoherent fluctuations plays the principal role in the case of transformation

* See also Farley et al. (3).

of electromagnetic waves with a small frequency change. For $B_0 \to 0$ formula (11.20) goes over into (10.40).

The ratio of the transformation coefficient (11.20) to the scattering coefficient (11.14) is c^3/s^3 in order of magnitude. Hence, in the domain of frequencies ω_0, near to $\omega_+(\vartheta_0)$ and $\omega_-(\vartheta_0)$, the absorption associated with electromagnetic wave transformation into Langmuir waves is more essential than the electromagnetic wave scattering.

3. Electromagnetic Wave Scattering by Langmuir Fluctuations

In addition to the principal maximum at $\Delta\omega = 0$ there are also maximums in the scattered radiation spectrum which are associated with electromagnetic wave scattering and transformation by the coherent collective fluctuations in a plasma. Let us first consider electromagnetic wave scattering by high-frequency Langmuir fluctuations. In this case interaction of the incident wave with the electron-density fluctuations plays the major role. Hence, the electromagnetic wave scattering cross section will be determined by (11.14).

If $aq \ll 1$, the correlation function (11.17) possesses delta-like maximums at frequencies corresponding to the Langmuir plasma oscillations in the magnetic field:

$$\omega_{\pm}^2 = \tfrac{1}{2}(\Omega^2 + \omega_B^2) \pm \tfrac{1}{2}[\Omega^2 + \omega_B^2)^2 - 4\Omega^2\omega_B^2 \cos^2 \tilde{\vartheta}]^{1/2}, \qquad (11.21)$$

where $\tilde{\vartheta}$ is the angle between $\mathbf{q}$ and $\mathbf{B}_0$, connected to the angles ϑ_0, ϑ, and φ by the relationship

$$\tan^2 \tilde{\vartheta} = \frac{k_0^3 \sin^2 \vartheta_0 + k^2 \sin^2 \vartheta - 2k_0 k \sin \vartheta_0 \sin \vartheta \cos \varphi}{(k_0 \cos \vartheta_0 - k \cos \vartheta)^2}. \qquad (11.22)$$

The scattering cross section for frequency shifts $\Delta\omega$, close to the Langmuir oscillation frequencies $\omega_\pm$, has the form

$$d\Sigma = \tfrac{1}{2}n_0 \left(\frac{e^2}{mc^2}\right)^2 \frac{\omega_0^2\omega^2 \,\Delta\omega^2}{\Omega^6} \, a^2 q^2 R \mid \xi \mid^2 \frac{(\omega_B^2 - \Delta\omega^2)^2}{\Delta\omega^4 \sin^2 \tilde{\vartheta} + (\omega_B^2 - \Delta\omega^2)^2 \cos^2 \tilde{\vartheta}}$$

$$\times \{\delta(\Delta\omega - \omega_+) + \delta(\Delta\omega + \omega_+) + \delta(\Delta\omega - \omega_-) + \delta(\Delta\omega + \omega_-)d\omega \, do.$$
$$(11.23)$$

In the case of a nonequilibrium plasma (a plasma permeated by a beam of charged particles, say), it is necessary to take account of the additional factor

$$\frac{R(\omega, \mathbf{k})}{|\, 1 - (u/\tilde{u}) \cos \tilde{\vartheta} \,|}$$

due to replacement of the temperature by the effective temperature (7.35) in the cross section (11.23). The relative contribution of the combined scattering (11.23) to the integral coefficient (11.19) is a quantity on the order of $a^2 q^2$. However, under nonequilibrium conditions the combined scattering cross section of electromagnetic waves may grow anomalously if the plasma is near the kinetic instability domain.

4. Electromagnetic Wave Scattering by Alfven and Magnetosonic Fluctuations

Combined scattering of electromagnetic waves in a magneto-active plasma is also possible by low-frequency magnetosonic and Alfven fluctuations. According to (5.87) the current inducing the scattered waves (11.5) may be expressed in the case of coherent fluctuations, only in terms of the electric field fluctuations. Utilizing the general expression for the scattering cross section (11.12) and the formulas (7.29)–(7.31) for the spectral distributions of the low-frequency fluctuations, various specific cases of wave scattering may easily be investigated.

In a nonisothermal plasma the electromagnetic wave scattering by slow magnetosonic fluctuations turns out to be most essential. In this case the density fluctuations play the major part in (11.5). The electromagnetic wave scattering cross section by slow magnetosonic fluctuations is determined by the expression

$$d\Sigma = \tfrac{1}{2} n_0 \left(\frac{e^2}{mc^2} \right)^2 \frac{\omega_0^4}{\Omega^4} \frac{\eta^3 \,|\, \mathbf{e}*(\hat{\epsilon} - 1)\mathbf{e}_0 \,|^2}{\eta_0 \left(|\, \mathbf{e}_0 \,|^2 - \dfrac{|\, \mathbf{k}_0 \mathbf{e}_0 \,|^2}{k_0^2} \right)} (\mathbf{e}*\hat{\epsilon}\mathbf{e})$$

$$\times \{\delta(\varDelta\omega - qv_s \cos \tilde{\vartheta}) + \delta(\varDelta\omega + qv_s \cos \tilde{\vartheta})\} \, d\omega \, do,$$

$$(11.24)$$

where $\mathbf{e}$ and η are the polarization vector and the refractive index of the scattered electromagnetic wave, respectively.

The cross section (11.24) differs from the corresponding cross section in an isotropic plasma (10.32) only in the form of the dispersion for the nonisothermal fluctuations. The ratio of the electromagnetic wave scattering cross section by slow magnetosonic fluctuations, integrated with respect to the frequencies, in a quite nonisothermal plasma to (11.19), is a quantity on the order of one. Hence, the principal line in the scattered radiation spectrum in a quite nonisothermal plasma is split into two lines associated with the scattering by slow magnetosonic fluctuations.

In the case of electromagnetic wave scattering by fast magnetosonic fluctuations, magnetic field fluctuations must be taken into account in addition to the density fluctuations. The corresponding cross section is then defined by

$$d\Sigma = \tfrac{1}{2} n_0 \left(\frac{e^2}{mc^2}\right)^2 \frac{\omega_0^4}{\Omega^4} \frac{v_s^2}{v_A^2} \frac{\eta^3 \, |\, \mathbf{e}^* \hat{Q}^M \mathbf{e}_0 \,|^2}{\eta_0 \left(|\, \mathbf{e}_0 \,|^2 - \dfrac{|\, \mathbf{k}_0 \mathbf{e}_0 \,|^2}{k_0^2}\right)(\mathbf{e}^* \hat{\epsilon} \mathbf{e})}$$

$$\times \{\delta(\Delta\omega - qv_A) + \delta(\Delta\omega + qv_A)\} \, d\omega \, do, \qquad (11.25)$$

$$Q_{ij}^M = (\epsilon_{ik} - \delta_{ik}) \left\{ \sin\tilde{\vartheta}\, \delta_{kj} + i\, \frac{\omega_0 \omega_B}{q\Omega^2} (q_k \tilde{e}_l - \tilde{e}_k q_l)(\epsilon_{lj}^0 - \delta_{lj}) \right\}.$$

The ratio of the scattering cross section (11.25), integrated with respect to the frequencies, to (11.19) is a quantity of the order of v_s^2/v_A^2.

In the case of electromagnetic wave scattering by Alfven fluctuations, the magnetic field fluctuations play the major part since Alfven oscillations are not accompanied by a density change. The electromagnetic wave scattering cross section by Alfven fluctuations is determined by the expression

$$d\Sigma = \tfrac{1}{6} n_0 \left(\frac{e^2}{mc^2}\right)^2 \frac{\omega_0^6}{\Omega^6} \frac{s^2}{c^2 \cos^2\tilde{\vartheta}} \frac{\eta^3 \, |\, \mathbf{e}^* \hat{Q}^A \mathbf{e}_0 \,|^2}{\eta_0 \left(|\, \mathbf{e}_0 \,|^2 - \dfrac{|\, \mathbf{k}_0 \mathbf{e}_0 \,|^2}{k_0^2}\right)(\mathbf{e}^* \hat{\epsilon} \mathbf{e})}$$

$$\times \{\delta(\Delta\omega - qv_A \cos\tilde{\vartheta}) + \delta(\Delta\omega + qv_A \cos\tilde{\vartheta})\} \, d\omega \, do, \qquad (11.26)$$

$$Q_{ij}^A = \frac{1}{q}(\epsilon_{ik} - \delta_{ik})(q_k \tilde{e}_l - \tilde{e}_k q_l)(\epsilon_{lj}^0 - \delta_{lj}).$$

Formula (11.26) is valid for both the isothermal and the non-isothermal cases.

The ratio of the electromagnetic wave scattering cross section by Alfven fluctuations, integrated with respect to the frequencies, to (11.19) is a quantity on the order of s^2/c^2 .

The electromagnetic wave scattering and transformation cross sections by low-frequency fluctuations may, exactly as in the Langmuir fluctuations case, increase radically in a nonequilibrium plasma if the latter is near the kinetic instability domain.

REFERENCES

1. A. I. Akhiezer, I. A. Akhiezer, and A. G. Sitenko, *ZETF* **41**, 644 (1961) *(Sov. Phys. JETP)*.
2. A. G. Sitenko and Iu. A. Kirochkin, *Izv. Vuz, Radiofizika* **6**, 469 (1963) *(CFSTI)*.
3. D. T. Farley, J. P. Dougherty, and D. W. Barron, *Proc. Roy. Soc. (London)* **A263**, 238 (1961).
4. E. E. Salpeter, *Phys. Rev.* **122**, 1663 (1961).

Quantum Plasma. Fluctuations in a Degenerate Electron Gas

1. Space-Time Correlations of Fluctuations in an Ideal Fermi Gas

Up to now we have limited ourselves to the analysis of the properties of a classical plasma. However, under definite conditions (for example, for electrons in metals) the quantum effects may play a significant part. The electrodynamic properties of a quantum plasma may also be investigated on the basis of inversion of the fluctuation-dissipation theorem. To do this, let us first consider the space-time correlations of fluctuations in an ideal Fermi gas.

The fluctuations in the density of the number of particles and the density of the current may be examined most simply by utilizing the second quantization method. Let us define the operator of the density of the number of particles by using the relationship

$$\hat{n}(\mathbf{r}, t) = \sum_{s=-\frac{1}{2}, \frac{1}{2}} \psi_s^+(\mathbf{r}, t)\, \psi_s(\mathbf{r}, t), \tag{12.1}$$

where ψ_s and ψ_s^+ are second-quantized wavefunctions

$$\psi_s(\mathbf{r}, t) = \frac{1}{\sqrt{V}} \sum_{\mathbf{p}} a_{\mathbf{p}s} e^{i(\mathbf{p}\mathbf{r} - E_{\mathbf{p}}t)}, \quad \psi_s^+(\mathbf{r}, t) = \frac{1}{\sqrt{V}} \sum_{\mathbf{p}} a_{\mathbf{p}s}^+ e^{-i(\mathbf{p}\mathbf{r} - E_{\mathbf{p}}t)}, \tag{12.2}$$

$a_{\mathbf{p}s}$ and $a_{\mathbf{p}s}^+$ are operators of the absorption and generation of particles in a state with momentum $\mathbf{p}$, and spin projection s, $E_\mathbf{p}$ is the energy of a particle with momentum $\mathbf{p}$ (we assume a quadratic dispersion law $E_\mathbf{p} = p^2/2m$, m is the particle mass), and V is the volume of the system ($\hbar = 1$). Substituting (12.2) into (12.1), we have

$$\hat{n}(\mathbf{r}, t) = \frac{1}{V} \sum_{\mathbf{p}\mathbf{p}'s} a^+_{\mathbf{p}s} a_{\mathbf{p}'s} e^{-i(\mathbf{p}-\mathbf{p}')\mathbf{r}+i(E_\mathbf{p}-E_{\mathbf{p}'})t}. \tag{12.3}$$

The diagonal terms ($\mathbf{p} = \mathbf{p}'$) of the operator $\hat{n}$ determine the mean particle density n_0. The difference between $\hat{n}$ and the mean value n_0 may be considered as the operator of the density fluctuations

$$\delta\hat{n}(\mathbf{r}, t) \equiv \hat{n}(\mathbf{r}, t) - n_0 . \tag{12.4}$$

The correlation function for the density fluctuations is determined by the expression

$$\langle \delta n^2 \rangle^0_{rt} = \langle \delta n(\mathbf{r}_1 , t_1) \, \delta n(\mathbf{r}_2 , t_2) \rangle, \tag{12.5}$$

where the average is taken with respect to the quantum-mechanical states of the system and with respect to the equilibrium distribution of the various quantum-mechanical states of the system.

Upon multiplying the operators (12.4) and (12.5) we obtain a sum of terms containing various products of the operators $a_{\mathbf{p}s}$ and $a^+_{\mathbf{p}s}$, taken in fours. The mean value of the products of the four Fermion operators a may be expressed in terms of the mean value of the products of the same operators, by two's:

$$\langle a_1 a_2 a_3 a_4 \rangle = \langle a_1 a_2 \rangle \langle a_3 a_4 \rangle - \langle a_1 a_3 \rangle a_2 a_4 \rangle + \langle a_1 a_4 \rangle \langle a_2 a_3 \rangle. \tag{12.6}$$

The mean value of the product of the two operators a_1 and a_2 differs from zero only if these operators are Hermitian conjugates. Let $n_\mathbf{p}$ denote the mean value of the product $a^+_{\mathbf{p}s} a_{\mathbf{p}s}$ determining the mean number of particles in a state with given values of $\mathbf{p}$ and s:

$$n_\mathbf{p} = \langle a^+_{\mathbf{p}s} a_{\mathbf{p}s} \rangle. \tag{12.7}$$

This number is independent of the value of the spin projection s because of the isotropy of the space.

The distribution function of a Fermi gas defines

$$n_{\mathbf{p}} = \left(\exp\left[\frac{E_{\mathbf{p}} - \mu}{T} \right] + 1 \right)^{-1}, \tag{12.8}$$

where μ is the chemical potential of the system.
It is easy to verify that

$$\langle a^+_{\mathbf{p}_1 s_1} a_{\mathbf{p}_1' s_1} a^+_{\mathbf{p}_2 s_2} a_{\mathbf{p}_2' s_2} \rangle = n_{\mathbf{p}_1} (1 - n_{\mathbf{p}_2}) \, \delta_{\mathbf{p}_1 \mathbf{p}_2'} \delta_{\mathbf{p}_1' \mathbf{p}_2} \delta_{s_1 s_2}, \tag{12.9}$$

and, hence,

$$\langle \delta n^2 \rangle^0_{\mathbf{r} t} = \frac{2}{V^2} \sum_{\mathbf{p}_1 \mathbf{p}_2} n_{\mathbf{p}_1} (1 - n_{\mathbf{p}_2}) \, e^{i(\mathbf{p}_1 - \mathbf{p}_2)\mathbf{r} - i(E_{\mathbf{p}_1} - F_{\mathbf{p}_2})t}. \tag{12.10}$$

Let us note that the correlation in an ideal Fermi gas is due to the Pauli principle, hence, it holds only between particles with identically directed spins. Utilizing (12.10), we obtain the following formula for the spectral distribution of the density fluctuations in an ideal Fermi gas:

$$\langle \delta n^2 \rangle^0_{\mathbf{k}\omega} = \frac{4\pi}{V} \sum_{\mathbf{p}} n_{\mathbf{p}} (1 - n_{\mathbf{p-k}}) \, \delta(\omega - E_{\mathbf{p}} + E_{\mathbf{p-k}}). \tag{12.11}$$

Remarking that

$$n_{\mathbf{p}}(1 - n_{\mathbf{p-k}}) \equiv (n_{\mathbf{p-k}} - n_{\mathbf{p}}) \left(\exp\left[\frac{E_{\mathbf{p}} - E_{\mathbf{p-k}}}{T} \right] - 1 \right)^{-1}, \tag{12.12}$$

the formula for the spectral distribution of the density fluctuations may be reduced to

$$\langle \delta n^2 \rangle^0_{\mathbf{k}\omega} = (e^{\omega/T} - 1)^{-1} \frac{4\pi}{V} \sum_{\mathbf{p}} n_{\mathbf{p}} \{ \delta(\omega - E_{\mathbf{p-k}} + E_{\mathbf{p}}) - \delta(\omega - E_{\mathbf{p}} + E_{\mathbf{p-k}}) \}. \tag{12.13}$$

In an analogous manner, by utilizing the customary definition of the particle-current operator

$$\mathbf{j}(\mathbf{r}, t) = \frac{i}{2m} \sum_{s} \{ \psi_s^+(\mathbf{r}, t) \, \nabla \psi_s(\mathbf{r}, t) - \nabla \psi_s^+(\mathbf{r}, t) \cdot \psi_s(\mathbf{r}, t) \}, \tag{12.14}$$

we obtain for the spectral distribution of the transverse current fluctuations

$$\langle \delta j_t^2 \rangle_{\mathbf{k}\omega}^0 = (e^{\omega/T} - 1)^{-1} \frac{4\pi}{V} \frac{1}{m^2} \sum_{\mathbf{p}} n_{\mathbf{p}} \left(p^2 - \frac{(\mathbf{k}\mathbf{p})^2}{k^2} \right) \{ \delta(\omega - E_{\mathbf{p}-\mathbf{k}} + E_{\mathbf{p}})$$

$$- \delta(\omega - E_{\mathbf{p}} + E_{\mathbf{p}-\mathbf{k}}) \}. \tag{12.15}$$

Considering the volume V to be sufficiently great, the summation in (12.13) and (12.15) may be replaced by integration

$$\sum_{\mathbf{p}} \cdots \to \frac{V}{(2\pi)^3} \int d\mathbf{p} \cdots .$$

We obtain the following formulas for the spectral distributions of the density and transverse-current fluctuations in an ideal Fermi gas for an arbitrary temperature T as a result of such integration:

$$\langle \delta n^2 \rangle_{\mathbf{k}\omega}^0 = (e^{\omega/T} - 1)^{-1} \frac{m^2 T}{\pi k} \ln \frac{1 + \exp\left[\frac{\mu}{T} - \left(\omega - \frac{k^2}{2m} \right)^2 \frac{m}{2k^2 T} \right]}{1 + \exp\left[\frac{\mu}{T} - \left(\omega + \frac{k^2}{2m} \right)^2 \frac{m}{2k^2 T} \right]},$$

$$\tag{12.16}$$

$$\langle \delta j_t^2 \rangle_{\mathbf{k}\omega}^0 = (e^{\omega/T} - 1)^{-1} \frac{2mT^2}{\pi k} \left\{ \zeta \left[\frac{\mu}{T} - \left(\omega - \frac{k^2}{2m} \right)^2 \frac{m}{2k^2 T} \right] \right.$$

$$\left. - \zeta \left[\frac{\mu}{T} - \left(\omega + \frac{k^2}{2m} \right)^2 \frac{m}{2k^2 T} \right] \right\}, \tag{12.17}$$

where

$$\zeta(x) = \int_0^\infty \frac{y\,dy}{e^{-x+y} + 1}.$$

In the zero temperature case ($T = 0$) the expressions for the spectral distributions of the fluctuations simplify substantially:

$$\langle \delta n^2 \rangle_{\mathbf{k}\omega} = \frac{3\pi}{4} \frac{n_0}{E_F} \begin{cases} \frac{p_F}{k} \frac{|\omega|}{E_F}, & -\frac{k}{p_F}\left(2 - \frac{k}{p_F}\right) < \frac{\omega}{E_F} < 0, \\[2mm] \frac{p_F}{k} \left\{ 1 - \frac{1}{4}\left(\frac{k}{p_F} + \frac{p_F}{k} \frac{\omega}{E_F} \right)^2 \right\}, \\[2mm] -\frac{k}{p_F}\left(\frac{k}{p_F} + 2 \right) < \frac{\omega}{E_F} < -\frac{k}{p_F}\left| \frac{k}{p_F} - 2 \right|; \end{cases} \tag{12.18}$$

$$\langle \delta j_t{}^2 \rangle^0_{k\omega} = \frac{3\pi}{4}\frac{n_0}{m} \begin{cases} \dfrac{p_F}{k}\dfrac{|\omega|}{E_F}\left\{2 - \dfrac{1}{4}\left(\dfrac{k}{p_F} + \dfrac{p_F}{k}\dfrac{\omega}{E_F}\right)^2 - \dfrac{1}{4}\left(\dfrac{k}{p_F} - \dfrac{p_F}{k}\dfrac{\omega}{E_F}\right)^2\right\}, \\[2ex] \qquad -\dfrac{k}{p_F}\left(2 - \dfrac{k}{p_F}\right) < \dfrac{\omega}{E_F} < 0, \\[2ex] \dfrac{p_F}{k}\left\{1 - \dfrac{1}{4}\left(\dfrac{k}{p_F} + \dfrac{p_F}{k}\dfrac{\omega}{E_F}\right)^2\right\}, \\[2ex] \qquad -\dfrac{k}{p_F}\left(\dfrac{k}{p_F} + 2\right) < \dfrac{\omega}{E_F} < -\dfrac{k}{p_F}\left|\dfrac{k}{p_F} - 2\right|, \end{cases}$$

(12.19)

where E_F is the boundary Fermi energy ($E_F = p_F{}^2/2m$). Let us note that at $T = 0$ fluctuations with negative frequencies are possible since the system of Fermi particles is in the ground state. Integrating (12.18) and (12.19) with respect to the frequencies, we may obtain the following formulas for the correlation functions of the simultaneous fluctuations in an ideal Fermi gas at $T = 0$:

$$\langle \delta n^2 \rangle_k{}^0 = n_0 \begin{cases} \dfrac{1}{4}\dfrac{k}{p_F}\left(3 - \dfrac{1}{4}\dfrac{k^2}{p_F{}^2}\right), & k < 2p_F, \\[2ex] 1, & k > 2p_F; \end{cases}$$

(12.20)

$$\langle \delta j_t{}^2 \rangle_k{}^0 = \frac{2}{5}n_0 v_F{}^2 \begin{cases} \dfrac{15}{16}\dfrac{k}{p_F}\left(1 - \dfrac{1}{6}\dfrac{k^2}{p_F{}^2} + \dfrac{1}{80}\dfrac{k^4}{p_F{}^4}\right), & k < 2p_F, \\[2ex] 1, & k > 2p_F. \end{cases}$$

(12.21)

The difference between $\langle \delta n^2 \rangle_k{}^0$ and n_0 at $k < 2p_F$ is due to the influence of the Pauli principle.

2. Longitudinal and Transverse Dielectric Permittivities of a Quantum Electron Gas

The obtained expressions for the spectral distributions of fluctuations in an ideal Fermi gas may be utilized to find the dielectric permittivities of a degenerate electron gas. Substituting

(12.13) into (2.64) and (12.15) into (2.65), we obtain the following expressions for the longitudinal and transverse dielectric permittivities of an electron gas [Lindhard (1)]:

$$\epsilon_l(\omega, \mathbf{k}) = 1 - \frac{8\pi e^2}{k^2} \frac{1}{V} \sum_{\mathbf{p}} n_{\mathbf{p}} \left\{ \frac{1}{\omega - E_{\mathbf{p-k}} + E_{\mathbf{p}} + io} - \frac{1}{\omega - E_{\mathbf{p}} + E_{\mathbf{p-k}} + io} \right\},$$

$$\tag{12.22}$$

$$\epsilon_t(\omega, \mathbf{k}) = 1 - \frac{\Omega^2}{\omega^2} - \frac{4\pi e^2}{\omega^2 m^2} \frac{1}{V} \sum_{\mathbf{p}} n_{\mathbf{p}} \left(p^2 - \frac{(\mathbf{kp})^2}{k^2} \right)$$

$$\times \left\{ \frac{1}{\omega - E_{\mathbf{p-k}} + E_{\mathbf{p}} + io} - \frac{1}{\omega - E_{\mathbf{p}} + E_{\mathbf{p-k}} + io} \right\}. \tag{12.23}$$

In the case of a completely degenerate electron gas $(T = 0)$ the dielectric permittivities are determined by the formulas:

$$\epsilon_l(\omega, \mathbf{k}) = 1 + \frac{3}{8} \frac{\Omega^2}{E_F^2} \frac{1}{q^2} \left\{ 1 + \frac{1}{2q} \left[1 - \frac{1}{4} \left(q - \frac{u}{q} \right)^2 \right] \ln \frac{q(q + 2) - u}{q(q - 2) - u} \right.$$

$$\left. + \frac{1}{2q} \left[1 - \frac{1}{4} \left(q + \frac{u}{q} \right)^2 \right] \ln \frac{q(q + 2) + u}{q(q - 2) + u} \right\}, \tag{12.24}$$

$$\epsilon_t(\omega, \mathbf{k}) = 1 + \frac{3}{8} \frac{\Omega^2}{E_F^2} \frac{1}{u^2} \left\{ 1 + \frac{1}{4} q^2 + \frac{3}{4} \frac{u^2}{q^2} \right.$$

$$- \frac{1}{2q} \left[1 - \frac{1}{4} \left(q - \frac{u}{q} \right)^2 \right]^2 \ln \frac{q(q + 2) - u}{q(q - 2) - u}$$

$$\left. - \frac{1}{2q} \left[1 - \frac{1}{4} \left(q + \frac{u}{q} \right)^2 \right]^2 \ln \frac{q(q + 2) + u}{q(q - 2) + u} \right\}, \tag{12.25}$$

where we have introduced the notation $u = |\omega|/E_F$ and $q = k/p_F$.

By using the expressions for ϵ_l and ϵ_t the magnetic permittivity of an electron gas may be determined:

$$1 - \frac{1}{\mu(\omega, \mathbf{k})} = \frac{\omega^2}{k^2 c^2} \{ \epsilon_t(\omega, \mathbf{k}) - \epsilon_l(\omega, \mathbf{k}) \}. \tag{12.26}$$

Thus in the limiting case of a static homogeneous field, we find by utilizing (12.22) and (12.23)

$$1 - \frac{1}{\mu(0, 0)} = \frac{4\pi e^2}{3m^2 c^2} \int \frac{d\mathbf{p}}{(2\pi)^3} \frac{\partial n_{\mathbf{p}}}{\partial \mu}. \tag{12.27}$$

Noting that the right-hand side of (12.27) is small as compared with one, (12.27) may be rewritten as

$$\chi(0, 0) = \frac{e^2}{3m^2c^2} \int \frac{d\mathbf{p}}{(2\pi)^3} \frac{\partial n_{\mathbf{p}}}{\partial \mu},$$

(12.28)

where $\chi(0, 0)$ is the static magnetic susceptibility of the electron gas. In the case of complete degeneration of the electron gas, the static magnetic susceptibility equals

$$\chi(0, 0) = \frac{e^2}{6\pi mc^2} \left(\frac{3n_0}{\pi} \right)^{1/3}.$$

(12.29)

Formula (12.29) takes account of Pauli spin paramagnetism and Landau orbital diamagnetism.

3. Dispersion of Plasma Oscillations

With the aid of the found expressions for ϵ_l and ϵ_t it is easy to determine the dispersion (dependence of ω on k) of the proper oscillations in a quantum electron gas. The dispersion of the longitudinal oscillations of an electron gas is determined from the condition

$$\epsilon_l(\omega, \mathbf{k}) = 0.$$

Plasma oscillations exist in the long wavelength range whose frequency is determined in the case of a completely degenerate electron gas by the expression [Klimontovich and Silin (2)]:

$$\omega^2 = \Omega^2 + \frac{3}{5} k^2 v_F^2 + \frac{k^4}{4m^2}.$$

(12.30)

The plasma oscillations correspond to the collective excitations in the spectrum of the electron gas. In the short wavelength range $(kv_F \gg \Omega)$ the spectrum of the longitudinal oscillations reduces to two-particle excitations of an electron-hole.

If the temperature differs from zero, the plasma oscillations are then characterized by weak attenuation. The attenuation coefficient is determined by the expression (3):

$$\gamma = \frac{m^3 \omega^3 T}{4\pi n_0 k^3} \ln \frac{1 + \exp\left[\frac{\mu}{T} - \left(\omega - \frac{k^2}{2m}\right)^2 \frac{m}{2k^2 T}\right]}{1 + \exp\left[\frac{\mu}{T} - \left(\omega + \frac{k^2}{2m}\right)^2 \frac{m}{2k^2 T}\right]}. \qquad (12.31)$$

At a zero temperature there is no attenuation of the plasma oscillations.

Dispersion of the transverse oscillations is determined from the condition

$$\epsilon_t(\omega, \mathbf{k}) - \eta^2 = 0.$$

Utilizing (12.25), the following expressions may be obtained for the square of the frequency of the transverse plasma oscillations:

$$\omega^2 = \Omega^2 + k^2 c^2 + \frac{1}{5} k^2 v_F^2, \qquad k^2 c^2 \ll \Omega^2,$$

$$(12.32)$$

$$\omega^2 = k^2 c^2 + \Omega^2 \left(1 + \frac{1}{5} \frac{v_F^2}{c^2}\right), \qquad k^2 c^2 \gg \Omega^2.$$

4. Fluctuations in a Degenerate Electron Gas

The obtained expressions for the dielectric permittivities (12.22) and (12.23) permit an investigation also of the density and current fluctuations in a degenerate electron gas taking account of Coulomb interaction between electrons. According to (2.50), the spectral distribution of the density fluctuations of the electron gas is determined by the expression

$$\langle \delta n^2 \rangle_{\mathbf{k}\omega} = \frac{k^2}{2\pi e^2} \frac{1}{e^{\omega/T} - 1} \frac{\text{Im } \epsilon_l(\omega, \mathbf{k})}{|\epsilon_l(\omega, \mathbf{k})|^2}, \qquad (12.33)$$

where (12.22) should be used for ϵ_l. The spectral distribution (12.33) is essentially different from the spectral distribution of the fluctuations for the ideal gas (12.16) only in the long wavelength portion of the spectrum in which collective effects associated with the Coulomb interaction between electrons are essential. In this range of the spectrum the spectral distribution $\langle\delta n^2\rangle_{k\omega}$, in contrast to $\langle\delta n^2\rangle_{k\omega}^0$, contains sharp maximums at frequencies corresponding to the proper oscillations of the degenerate electron gas.

Let us note that in deriving formula (12.33) on the basis of perturbation theory it turns out to be necessary to take account of an infinite series of diagrams describing the shielding of the charge in the plasma (4).*

Let us examine the correlation function of the electron–gas density in the case of complete degeneration ($T = 0$) in more detail. If $T = 0$, we have for the spectral distribution of the correlation function in the negative frequency range ($\omega < 0$) (3):

$$\langle\delta n^2\rangle_{k\omega} = \frac{k^2}{2\pi e^2} \frac{\epsilon_l''}{\epsilon_l'^2 + \epsilon_l''^2},\tag{12.34}$$

where ϵ_l' and ϵ_l'' are the real and imaginary parts of the longitudinal dielectric permittivity ϵ_l, defined by the equalities:

$$\epsilon_l' = 1 + \frac{3}{8}\frac{\Omega^2}{E_F^2}\frac{1}{q^2}\left\{1 + \frac{1}{2q}\left[1 - \frac{1}{4}\left(q - \frac{u}{q}\right)^2\right]\ln\frac{|\,q(q+2)-u\,|}{|\,q(q-2)-u\,|}\right.$$
$$\left.+ \frac{1}{2q}\left[1 - \frac{1}{4}\left(q + \frac{u}{q}\right)^2\right]\ln\frac{|\,q(q+2)+u\,|}{|\,q(q-2)+u\,|}\right\},\tag{12.35}$$

$$\epsilon_l'' = \frac{3\pi}{16}\frac{\Omega^2}{E_F^2}\frac{1}{q^3}\begin{cases} u, & 0 < u < q(2-q), \\ 1 - \frac{1}{4}\left(q - \frac{u}{q}\right)^2, & q\,|\,q-2\,| < u < q(q+2), \\ 0, & 0 < u < q(q-2), \quad u > q(q+2). \end{cases}\tag{12.36}$$

Let us recall that $u = |\,\omega\,|/E_F$ and $q = k/p_F$.

Domains for different values of ϵ_l'' are shown in Fig. 22. Since ϵ_l'' vanishes for $u > q(q+2)$ (domain 3), density fluctuations are then possible in this range of ω and k values only at frequencies satisfying the condition $\epsilon_l'(\omega, \mathbf{k}) = 0$ and agreeing with the proper frequencies of the undamped plasma oscillations. For $0 < u < q(2-q)$, $q(q-2) < u < q(q+2)$ (1 and 2), the density fluctuations are characterized by a continuous spectrum and are connected

* Akhiezer (5) considered the properties of a relativistic degenerate electron gas.

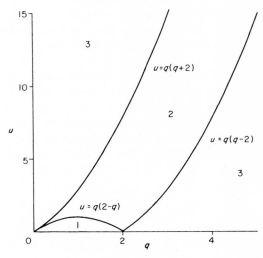

FIG. 22.

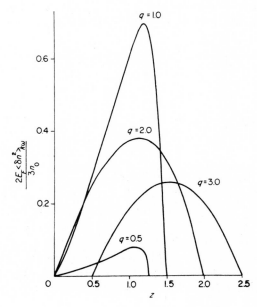

FIG. 23.

with two-particle (electron-hole) excitations in the system. The curves $u = q(q + 2)$ and $u = q(q - 2) > 0$ have a simple physical meaning; for a fixed value of k they bound the spectrum of two-particle excitations.

Evidently, undamped plasma oscillations are possible only if the plasma frequency lies above the domain of the continuous spectrum for a fixed k. For a given electron density, the maximum value k_m for the plasma excitations is determined by the point of intersection of the curve $\epsilon'(u, q) = 0$, with the upper boundary of the continuous spectrum $u = q(q + 2)$:

$$\frac{1}{q_m{}^2} \left[(2 + q_m) \ln \left(1 + \frac{2}{q_m} \right) - 2 \right] = \frac{16}{3} \frac{E_F{}^2}{\Omega^2}, \qquad q_m = \frac{k_m}{p_F}. \tag{12.37}$$

The spectral distribution of the density fluctuations of a completely degenerate electron gas in shown in Fig. 23 as a function of $z = |\omega|/kv_F$ for different values of $q = 0.5$, 1, 2, and 3 $[\frac{3}{4}(\Omega^2/E_F{}^2) = 1]$.

In the long wavelength range $(q \ll 1)$, the spectral distribution of the density fluctuations may be represented as

$$\langle \delta n^2 \rangle_{\mathbf{k}\omega} = 3\pi \frac{n_0}{E_F} \left\{ \frac{1}{2} z\theta(1 - z) \Big/ \left(\left[1 + p \left(1 - \frac{z}{2} \ln \frac{1 + z}{1 - z} \right) \right]^2 + \frac{\pi^2}{4} p^2 z^2 \right) \right.$$
$$\left. + p^{-1}\delta \left[1 + p \left(1 - \frac{z}{2} \ln \frac{z + 1}{z - 1} \right) \right] \right\}, \tag{12.38}$$

$$p = \frac{3\Omega^2}{k^2 v_F{}^2}.$$

It is convenient to use the integral relationship

$$\int_0^\infty \mathrm{Im} \frac{1}{\epsilon_l(\omega, \mathbf{k})} \, d\omega = \int_0^\infty \left\{ \frac{1}{\epsilon_l(i\omega, \mathbf{k})} - 1 \right\} d\omega \tag{12.39}$$

for the calculation of $\langle \delta n^2 \rangle_{\mathbf{k}}$. Use of this relationship yields

$$\langle \delta n^2 \rangle_{\mathbf{k}} = \frac{3}{4} q\eta(p) \, n_0, \qquad \eta(p) \equiv \frac{4}{\pi} \int_0^\infty \frac{1 - z \arctan 1/z}{1 + p(1 - z \arctan 1/z)} \, dz. \tag{12.40}$$

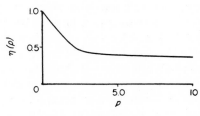

FIG. 24.

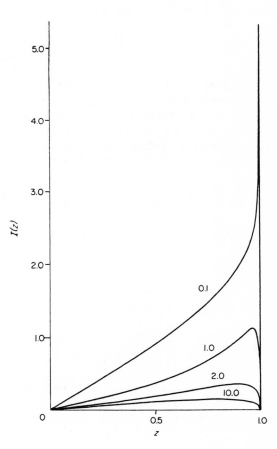

FIG. 25.

The function $\eta(p)$ characterizes the shielding of the Coulomb charge in a degenerate electron gas. The dependence of the function $\eta(p)$ on p is shown in Fig. 24.

The dependence of the spectral density of the fluctuations $I(z) \equiv 2\langle \delta n^2 \rangle_{kz}/\langle \delta n^2 \rangle_k$ on z in the $(0, 1)$ range is shown in Fig. 25 for different values of the parameter p $(p = 0.1; 1; 2;$ and $10)$. In the $p = 10$ case the intensity $I(z)$ is magnified fivefold.

Let us note that, in contrast to the classical case, $I(z)$ vanishes for $z = 0$, and this is related to the Fermi character of the electron velocity distribution. For $z > 1$ the function $I(z)$ has a delta-like maximum corresponding to plasma fluctuations of the degenerate electron gas. The integral $w(p) \equiv \int_1^\infty I(z)\,dz$ determines the relative weight of the density fluctuations at the plasma frequency. (The function $I(z)$ is normalized according to the condition $\int_0^\infty I(z)\,dz = 1$.) The value of this integral as a function of the parameter p, characterizing the intensity of interaction between electrons is shown in Fig. 26. Let us note that as the parameter $p = 3\Omega^2/k^2 v_F^2$ increases the weight of the low-frequency fluctuations, associated with the two-particle excitations, decreases.

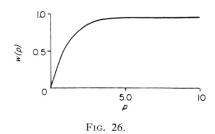

FIG. 26.

REFERENCES

1. J. Lindhard, *Dan. Mat. Fys. Medd.* **28**(8) (1954).
2. Iu. L. Klimontovich and V. P. Silin, *ZETF* **23**, 151 (1952) (CCT).
3. A. G. Sitenko and I. V. Simenog, *Izv. Vuz, Radiofizika* **6**, 54 (1963) (CFSTI).
4. A. J. Glick, *Ann. Phys.* **17**, 61 (1962).
5. I. A. Akhiezer, *Ukr. Fiz. Zh.* **6**, 435 (1961).

13

Fluctuations in a
Superconducting Plasma

---

1. Superconducting State in a System of Electrons

Electromagnetic fluctuations in a system of electrons in the super-conducting state are characterized by a number of particularities associated with the structure of the excitation spectrum in such a system. The microscopic theory of superconductivity has been developed by Bardeen, Cooper, and Schrieffer (1), and Bogoliubov (2) [see also Bogoliubov, Tolmachev, and Shirkov (3)]. On the basis of the microscopic theory of superconductivity (particularly, the Bogoliubov quasi-particle transformation), it is easy to determine the correlation functions for the density and current fluctuations of a system of electrons in the superconducting state, which may be used to investigate the electromagnetic properties of a super-conductor.*

Let us consider a system of electrons described by the simplified Bardeen Hamiltonian:

$$H = \sum_{\mathbf{p}s} \xi_{\mathbf{p}} a_{\mathbf{p}s}^{+} a_{\mathbf{p}s} - \frac{1}{V} \sum_{\substack{\mathbf{p} \neq \mathbf{p}' \\ s}} I(p, p') a_{\mathbf{p}s}^{+} a_{-\mathbf{p}-s}^{+} a_{-\mathbf{p}'-s} a_{\mathbf{p}'s} , \qquad (13.1)$$

where $a_{\mathbf{p}s}^{+}$ and $a_{\mathbf{p}s}$ are the generation and absorption operators of

---

* Abrikosov, Gor'kov, and Khalatnikov (4), Mattis and Bardeen (5), and Rickaysen (6) analyzed the electromagnetic properties of superconductors.

an electron in a state with momentum $\mathbf{p}$ and spin projection s, $\xi_{\mathbf{p}}$ is the energy of the electron measured from the Fermi surface, V the volume of the system, and $I(p, p')$ the interaction between electrons near the Fermi surface due to electron–phonon inter-action. It is assumed in (13.1) that the interaction takes place only between electrons in which the momentums and spins are oppositely directed.

By using the Bogoliubov quasi-particle transformation

$$a_{2sps} = u_{\mathbf{p}}\alpha_{\mathbf{p}-s} + 2sv_{\mathbf{p}}\alpha_{\mathbf{p}s}^{+} , \tag{13.2}$$

$$u_{\mathbf{p}}^{2} = \frac{1}{2}\left(1 + \frac{\xi_{\mathbf{p}}}{E_{\mathbf{p}}}\right), \qquad v_{\mathbf{p}}^{2} = \frac{1}{2}\left(1 - \frac{\xi_{\mathbf{p}}}{E_{\mathbf{p}}}\right), \tag{13.3}$$

the Hamiltonian (13.2) may be reduced to diagonal form

$$H = U + \sum_{ps} E_{\mathbf{p}}\alpha_{\mathbf{p}s}^{+}\alpha_{\mathbf{p}s} , \tag{13.4}$$

where U is the energy of a quasi-particle vacuum, $E_{\mathbf{p}} = (\xi_{\mathbf{p}}^{2} + \varDelta_{\mathbf{p}}^{2})^{1/2}$ is the energy of a quasi-particle with momentum $\mathbf{p}$ and gap width $\varDelta_{\mathbf{p}}$ in the quasi-particle spectrum determined from the equation

$$\varDelta_{\mathbf{p}} = \frac{1}{V}\sum_{\mathbf{p}'}\frac{\varDelta_{\mathbf{p}'}}{E_{\mathbf{p}'}}I(p, p') \tanh\frac{E_{\mathbf{p}'}}{2T} . \tag{13.5}$$

(We neglect interaction between quasi-particles.) According to (13.4), the following quadratic expressions of the quasi-particle operators:

$$\langle\alpha_{\mathbf{p}s}^{+}(t_{1})\,\alpha_{\mathbf{p}s}(t_{2})\rangle = N_{\mathbf{p}}e^{-iE_{\mathbf{p}}(t_{2}-t_{1})},$$

$$\langle\alpha_{\mathbf{p}s}(t_{1})\,\alpha_{\mathbf{p}s}^{+}(t_{2})\rangle = (1 - N_{\mathbf{p}})\,e^{iE_{\mathbf{p}}(t_{2}-t_{1})}, \tag{13.6}$$

where $N_{\mathbf{p}}$ is the mean number of quasi-particles in a state with momentum determined by the Fermi distribution function

$$N_{\mathbf{p}} = \frac{1}{e^{E_{\mathbf{p}}/T} + 1} , \tag{13.7}$$

have mean values different from zero. The mean number of electrons in a state with a definite momentum $n_{\mathbf{p}}$ is expressed in terms of the mean number of quasi-particles N_p by means of the relationship

$$n_p = u_p^2 N_p + v_p^2(1 - N_p). \tag{13.8}$$

Utilizing (13.3) and (13.7), we find for the electron distribution in the superconducting state

$$n_p = \frac{1}{2}\left(1 - \frac{\xi_p}{E_p}\tanh\frac{E_p}{2T}\right). \tag{13.9}$$

2. Density Fluctuations in a Superconductor

The correlation function for the density fluctuations expressed in terms of electron generation and absorption operators in the Heisenberg representation is determined by the formula

$$\langle \delta n^2 \rangle_{rt} = \frac{1}{V^2} \sum_{\substack{p_1 \neq p_1', \ p_2 \neq p_2' \\ s_1, \ s_2}} \langle a_{p_1 s_1}^+(t_1)\, a_{p_1' s_1}(t_1)\, a_{p_2 s_2}^+(t_2)\, a_{p_2' s_2}(t_2) \rangle$$

$$\times \exp\left[-i(p_1 - p_1')\, r_1 - i(p_2 - p_2')\, r_2\right]. \tag{13.10}$$

Substituting (13.2) into (13.10) and taking account of (13.6), we easily find the space-time correlation function $\langle \delta n^2 \rangle_{rt}$ for a system of electrons in the superconducting state. Let us note that in the superconducting state the correlation holds between electrons with both parallel and antiparallel spins

$$\langle \delta n^2 \rangle_{rt} = \langle \delta n^2 \rangle_{rt}^{\uparrow\uparrow} + \langle \delta n^2 \rangle_{rt}^{\uparrow\downarrow}, \tag{13.11}$$

$$\langle \delta n^2 \rangle_{rt}^{\uparrow\uparrow} = \frac{2}{V^2} \sum_{p \neq p'} \{u_p^2 N_p e^{-iE_p t} + v_p^2(1 - N_p)\, e^{iE_p t}\}$$

$$\times \{u_{p'}^2(1 - N_{p'})\, e^{iE_{p'}t} + v_{p'}^2 N_{p'} e^{-iE_{p'}t}\} e^{i(p-p')r}, \tag{13.12}$$

$$\langle \delta n^2 \rangle_{rt}^{\uparrow\downarrow} = \frac{2}{V^2} \sum_{p \neq p'} u_p v_p u_{p'} v_{p'} \{N_p e^{-iE_p t} - (1 - N_p)\, e^{iE_p t}\}$$

$$\times \{N_{p'} e^{-iE_{p'}t} - (1 - N_{p'})\, e^{iE_{p'}t}\}\, e^{i(p-p')r}. \tag{13.13}$$

The spectral distributions of the correlation functions for electrons with parallel and antiparallel spins are determined by the formulas:

$$\langle \delta n^2 \rangle_{\mathbf{k}\omega}^{\uparrow\uparrow} = \frac{4\pi}{V} \sum_{\mathbf{p}} \{ u_{\mathbf{p}}{}^2 u_{\mathbf{p}-\mathbf{k}}^2 N_{\mathbf{p}}(1 - N_{\mathbf{p}-\mathbf{k}}) \, \delta(\omega - E_{\mathbf{p}} + E_{\mathbf{p}-\mathbf{k}})$$

$$+ \, v_{\mathbf{p}}{}^2 v_{\mathbf{p}-\mathbf{k}}^2 (1 - N_{\mathbf{p}}) \, N_{\mathbf{p}-\mathbf{k}} \delta(\omega - E_{\mathbf{p}-\mathbf{k}} + E_{\mathbf{p}})$$

$$+ \, u_{\mathbf{p}}{}^2 v_{\mathbf{p}-\mathbf{k}}^2 N_{\mathbf{p}} N_{\mathbf{p}-\mathbf{k}} \delta(\omega - E_{\mathbf{p}} - E_{\mathbf{p}-\mathbf{k}})$$

$$+ \, v_{\mathbf{p}}{}^2 u_{\mathbf{p}-\mathbf{k}}^2 (1 - N_{\mathbf{p}})(1 - N_{\mathbf{p}-\mathbf{k}}) \, \delta(\omega + E_{\mathbf{p}} + E_{\mathbf{p}-\mathbf{k}}) \}, \qquad (13.14)$$

$$\langle \delta n^2 \rangle_{\mathbf{k}\omega}^{\uparrow\downarrow} = \frac{4\pi}{V} \sum_{\mathbf{p}} u_{\mathbf{p}} v_{\mathbf{p}} u_{\mathbf{p}-\mathbf{k}} v_{\mathbf{p}-\mathbf{k}} \{ -N_{\mathbf{p}}(1 - N_{\mathbf{p}-\mathbf{k}}) \, \delta(\omega - E_{\mathbf{p}} + E_{\mathbf{p}-\mathbf{k}})$$

$$- \, (1 - N_{\mathbf{p}}) \, N_{\mathbf{p}-\mathbf{k}} \delta(\omega - E_{\mathbf{p}-\mathbf{k}} + E_{\mathbf{p}})$$

$$+ \, N_{\mathbf{p}} N_{\mathbf{p}-\mathbf{k}} \delta(\omega - E_{\mathbf{p}} - E_{\mathbf{p}-\mathbf{k}})$$

$$+ \, (1 - N_{\mathbf{p}})(1 - N_{\mathbf{p}-\mathbf{k}}) \, \delta(\omega + E_{\mathbf{p}} + E_{\mathbf{p}-\mathbf{k}}) \}. \qquad (13.15)$$

The spectral density of the total correlation function $\langle \delta n^2 \rangle_{\mathbf{k}\omega}$ is the sum of (13.14) and (13.15). Using relationships of the type (12.12) for the product of the occupation numbers of the quasi-particles $N_{\mathbf{p}}$ and $N_{\mathbf{p}-\mathbf{k}}$, the spectral distribution of the total correlation function for a system of electrons in the superconducting state at an arbitrary temperature may be represented as

$$\langle \delta n^2 \rangle_{\mathbf{k}\omega} = (e^{\omega/T} - 1)^{-1} \frac{2\pi}{V} \sum_{\mathbf{p}} \{ (N_{\mathbf{p}} - N_{\mathbf{p}-\mathbf{k}})(u_{\mathbf{p}} u_{\mathbf{p}-\mathbf{k}} - v_{\mathbf{p}} v_{\mathbf{p}-\mathbf{k}})^2$$

$$\times [\delta(\omega - E_{\mathbf{p}-\mathbf{k}} - E_{\mathbf{p}}) - \delta(\omega - E_{\mathbf{p}} + E_{\mathbf{p}-\mathbf{k}})]$$

$$+ \, (1 - N_{\mathbf{p}} - N_{\mathbf{p}-\mathbf{k}})(u_{\mathbf{p}} v_{\mathbf{p}-\mathbf{k}} + v_{\mathbf{p}} u_{\mathbf{p}-\mathbf{k}})^2$$

$$\times [\delta(\omega - E_{\mathbf{p}} - E_{\mathbf{p}-\mathbf{k}}) - \delta(\omega + E_{\mathbf{p}} + E_{\mathbf{p}-\mathbf{k}})] \}, \qquad (13.16)$$

$$(u_{\mathbf{p}} u_{\mathbf{p}-\mathbf{k}} - v_{\mathbf{p}} v_{\mathbf{p}-\mathbf{k}})^2 = \frac{1}{2} \left(1 + \frac{\xi_{\mathbf{p}} \xi_{\mathbf{p}-\mathbf{k}} - \Delta^2}{E_{\mathbf{p}} E_{\mathbf{p}-\mathbf{k}}} \right),$$

$$(u_{\mathbf{p}} v_{\mathbf{p}-\mathbf{k}} + v_{\mathbf{p}} u_{\mathbf{p}-\mathbf{k}})^2 = \frac{1}{2} \left(1 - \frac{\xi_{\mathbf{p}} \xi_{\mathbf{p}-\mathbf{k}} - \Delta^2}{E_{\mathbf{p}} E_{\mathbf{p}-\mathbf{k}}} \right). \qquad (13.17)$$

In the zero temperature case ($T = 0$), the expressions for the spectral distributions of the correlation functions simplify substantially. Thus, for the spectral density of the total correlation function at $T = 0$, we have

$$\langle \delta n^2 \rangle_{\mathbf{k}\omega} = \frac{1}{8\pi^2} \int d\mathbf{p} \left\{ 1 - \frac{\xi_\mathbf{p} \xi_{\mathbf{p}-\mathbf{k}} - \Delta^2}{E_\mathbf{p} E_{\mathbf{p}-\mathbf{k}}} \right\} \delta(\omega + E_\mathbf{p} + E_{\mathbf{p}-\mathbf{k}}). \quad (13.18)$$

Let us note that only negative frequencies are contained in the fluctuation spectrum at zero temperature. Since the quasi-particle energy always exceeds the gap width Δ, then $\langle \delta n^2 \rangle_{\mathbf{k}\omega}$ will differ from zero for the superconducting state only for $\omega < -2\Delta$.

Assuming that the dispersion law for the electrons is quadratic, it is convenient to introduce new integration variables $\xi = (p^2/2m) - E_F$ and $\xi' = [(\mathbf{p} - \mathbf{k})^2/2m] - E_F$ in (13.18):

$$\langle \delta n^2 \rangle_{\mathbf{k}\omega} = \frac{m^2}{4\pi k} \int_{-E_F}^{\infty} d\xi \int_{\xi - 2k \left(\frac{\xi + E_F}{2m} \right)^{1/2} + \frac{k^2}{2m}}^{\xi + 2k \left(\frac{\xi + E_F}{2m} \right)^{1/2} + \frac{k^2}{2m}} d\xi' \left(1 - \frac{\xi \xi' - \Delta^2}{EE'} \right)$$

$$\times \delta(\omega + E + E'). \quad (13.19)$$

Because of the delta function in the integrand of (13.19), the integration may only be carried out in the domains ξ and ξ', lying within the curve

$$\omega + (\xi^2 + \Delta^2)^{1/2} + (\xi'^2 + \Delta^2)^{1/2} = 0. \quad (13.20)$$

[Let us note that the curve (13.20) is symmetric relative to the point $\xi = \xi' = 0$.] If the domain bounded by the curve (13.20) does not emerge beyond the limits of integration in (13.19) (as is easy to verify, this occurs if $\omega(\omega + 2\Delta) < E_F^2$ and $k(k - 2p_F) < -2m[\omega(\omega + 2\Delta)]^{1/2}$), then the integration in (13.19) may be performed explicitly. In fact, the limits of integration in (13.19) may be extended to infinity in this case; hence, the term with $\xi \xi'$ vanishes from the integrand. Now turning to integration with respect to dE and dE', we find

$$\langle \delta n^2 \rangle_{\mathbf{k}\omega} = \frac{m^2}{\pi k} \int_{\Delta}^{|\omega| - \Delta} dE \, \frac{E (|\omega| - E) + \Delta^2}{\{(E^2 - \Delta^2)[(\omega + E)^2 - \Delta^2]\}^{1/2}}. \quad (13.21)$$

This expression may be reduced to

$$\langle \delta n^2 \rangle_{\mathbf{k}\omega} = \frac{2\Delta m^2}{\pi k} \left\{ (s+1)E\left(\frac{s-1}{s+1}\right) - \frac{2s}{s+1} K\left(\frac{s-1}{s+1}\right) \right\}, \quad s = \frac{|\omega|}{2\Delta},$$

$$(13.22)$$

where $E(x)$ and $K(x)$ are the complete elliptic integrals. Let us emphasize that (13.22) is valid for values of ω and k, satisfying the conditions

$$\omega(\omega + 2\Delta) < E_F^2,$$

$$p_F - \{p_F^2 - 2m[\omega(\omega + 2\Delta)]^{1/2}\}^{1/2} < k < p_F + \{p_F^2 - 2m[\omega(\omega + 2\Delta)]^{1/2}\}^{1/2}.$$

Let us note that for $k \gg p_F$ the function $\langle \delta n^2 \rangle_{\mathbf{k}\omega}$ decreases more rapidly than k^{-2} as k increases. The spectral density of the time correlation function at $T = 0$ is determined by the expression

$$\langle \delta n^2 \rangle_\omega = \frac{1}{4\pi^3} \int_0^\infty dp\,p^2 \int_0^\infty dp'\,p'^2 \left\{ 1 - \frac{\xi_p \xi_{p'} - \Delta^2}{E_p E_{p'}} \right\} \delta(\omega + E_p + E_{p'}).$$

$$(13.23)$$

At low frequencies $\omega(\omega + 2\Delta) \ll E_F^2$, only domains of integration near the Fermi surface are essential in (13.23). In this case the spectral density (13.23) may be expressed in terms of elliptic integrals

$$\langle \delta n^2 \rangle_\omega = \frac{2\Delta m^2 p_F^2}{\pi^3} \left\{ (s+1)E\left(\frac{s-1}{s+1}\right) - \frac{2s}{s+1} K\left(\frac{s-1}{s+1}\right) \right\}, \quad s = \frac{|\omega|}{2\Delta}.$$

$$(13.24)$$

The spectral density of the space correlation function for electrons with parallel and antiparallel spins at $T = 0$ is determined by the formulas:

$$\langle \delta n^2 \rangle_{\mathbf{k}}^{\uparrow\uparrow} = \frac{\Delta p_F m}{4\pi^2} \varphi\left(\frac{k}{p_F'}, \frac{\Delta}{E_F}\right), \qquad \langle \delta n^2 \rangle_{\mathbf{k}}^{\uparrow\downarrow} = \frac{\Delta p_F m}{4\pi^2} f\left(\frac{k}{p_F}, \frac{\Delta}{E_F}\right),$$

$$\varphi(q, p) = \frac{1}{2pq} \int_0^\infty dt\, t \left\{ 4qt - \frac{t^2 - 1}{[(t^2 - 1)^2 + p^2]^{1/2}} [([(t + q)^2 - 1]^2 + p^2)^{1/2} \right.$$

$$\left. - ([(t - q)^2 - 1]^2 + p^2)^{1/2}] \right\},$$

$$(13.25)$$

$$f(q, p) = \frac{p}{2q} \int_0^\infty dt \frac{t}{[(t^2 - 1)^2 + p^2]^{1/2}} \left\{ \text{arcsinh} \frac{(t + q)^2 - 1}{p} \right.$$
$$\left. - \text{arcsinh} \frac{(t - q)^2 - 1}{p} \right\}. \tag{13.25}$$

If $k = 0$, then

$$\varphi(0, p) = f(0, p) = \frac{\pi}{2} \frac{p}{\sqrt{1 + p^2}} \csc \frac{1}{2} \arctan p;$$

for $p \ll 1$ and $q \ll p$ we have

$$f \cong \pi \left(1 + \frac{1}{4} \frac{q^2}{p} \right),$$

i.e., as q grows the correlation of the electrons with antiparallel spins and the correlation of the electrons with parallel spins increase (as for the normal case).

3. Density Fluctuations Associated with Collective Excitations

The analysis based on the utilization of the Hamiltonian (13.1) does not take account of the collective excitations in the superconducting systems. However, collective excitations play an essential part for small values of k. Describing the system of electrons by the Hamiltonian

$$H = \sum_{ps} \xi_p a_{ps}^+ a_{ps} - \frac{1}{V} \sum_{\substack{p_1 p_2 p_1' p_2' s_1 s_2 \\ p_1 + p_2 = p_1' + p_2' \\ p_1 \neq p_1'}} I(\mathbf{p}_1, \mathbf{p}_2, \mathbf{p}_1', \mathbf{p}_2') a_{p_1 s_1}^+ a_{p_2 s_2}^+ a_{p_2' s_2} a_{p_1' s_1} \tag{13.26}$$

and taking account of electron interaction only near the Fermi surface, by using chain equations for the Green's function, we may obtain the following expression for the correlation function of the density fluctuations (7, 8):

$$\langle \delta n^2 \rangle_{k\omega} = \frac{2}{I} \frac{1}{e^{\omega/T} - 1} \text{Im} \frac{\theta^2 - (b - 1) d}{bd - \theta^2}, \tag{13.27}$$

where I is the interaction on the Fermi surface, and

$$b = 1 + \frac{1}{2V} \sum_{\mathbf{p}} \left\{ (N_{\mathbf{p}} - N_{\mathbf{p}-\mathbf{k}}) \left(1 + \frac{\xi_{\mathbf{p}}\xi_{\mathbf{p}-\mathbf{k}} - \varDelta^2}{E_{\mathbf{p}}E_{\mathbf{p}-\mathbf{k}}} \right) \right.$$

$$\times \left(\frac{1}{\omega - E_{\mathbf{p}-\mathbf{k}} + E_{\mathbf{p}} + io} - \frac{1}{\omega - E_{\mathbf{p}} + E_{\mathbf{p}-\mathbf{k}} + io} \right)$$

$$+ (1 - N_{\mathbf{p}} - N_{\mathbf{p}-\mathbf{k}}) \left(1 - \frac{\xi_{\mathbf{p}}\xi_{\mathbf{p}-\mathbf{k}} - \varDelta^2}{E_{\mathbf{p}}E_{\mathbf{p}-\mathbf{k}}} \right)$$

$$\left. \times \left(\frac{1}{\omega - E_{\mathbf{p}} - E_{\mathbf{p}-\mathbf{k}} + io} - \frac{1}{\omega + E_{\mathbf{p}} + E_{\mathbf{p}-\mathbf{k}} + io} \right) \right\},$$

$$d = 1 + \frac{I}{2V} \sum_{\mathbf{p}} \left\{ (N_{\mathbf{p}} - N_{\mathbf{p}-\mathbf{k}}) \left(1 - \frac{\xi_{\mathbf{p}}\xi_{\mathbf{p}-\mathbf{k}} + \varDelta^2}{E_{\mathbf{p}}E_{\mathbf{p}-\mathbf{k}}} \right) \right.$$

$$\times \left(\frac{1}{\omega - E_{\mathbf{p}-\mathbf{k}} + E_{\mathbf{p}} + io} - \frac{1}{\omega - E_{\mathbf{p}} + E_{\mathbf{p}-\mathbf{k}} + io} \right) \qquad (13.28)$$

$$+ (1 - N_{\mathbf{p}} - N_{\mathbf{p}-\mathbf{k}}) \left(1 + \frac{\xi_{\mathbf{p}}\xi_{\mathbf{p}-\mathbf{k}} + \varDelta^2}{E_{\mathbf{p}}E_{\mathbf{p}-\mathbf{k}}} \right)$$

$$\left. \times \left(\frac{1}{\omega - E_{\mathbf{p}} - E_{\mathbf{p}-\mathbf{k}} + io} - \frac{1}{\omega + E_{\mathbf{p}} + E_{\mathbf{p}-\mathbf{k}} + io} \right) \right\},$$

$$\theta = \frac{I}{2V} \sum_{\mathbf{p}} \frac{\omega \varDelta}{E_{\mathbf{p}}E_{\mathbf{p}-\mathbf{k}}} \left\{ (N_{\mathbf{p}} - N_{\mathbf{p}-\mathbf{k}}) \left(\frac{1}{\omega - E_{\mathbf{p}-\mathbf{k}} + E_{\mathbf{p}} + io} \right. \right.$$

$$\left. - \frac{1}{\omega - E_{\mathbf{p}} + E_{\mathbf{p}-\mathbf{k}} + io} \right) + (1 - N_{\mathbf{p}} - N_{\mathbf{p}-\mathbf{k}})$$

$$\left. \times \left(\frac{1}{\omega - E_{\mathbf{p}} - E_{\mathbf{p}-\mathbf{k}} + io} - \frac{1}{\omega + E_{\mathbf{p}} + E_{\mathbf{p}-\mathbf{k}} + io} \right) \right\}.$$

Let us limit ourselves to the consideration of the zero temperature case. We have $\mathrm{Im}\, b = \mathrm{Im}\, d = \mathrm{Im}\, \theta = 0$ in the $|\omega| < 2\varDelta$ frequency range, and therefore,

$$\langle \delta n^2 \rangle_{\mathbf{k}\omega} = \frac{2\pi d}{I} \, \delta(bd - \theta^2), \qquad -2\varDelta < \omega < 0. \qquad (13.29)$$

The condition

$$bd - \theta^2 = 0 \qquad (13.30)$$

determines the dispersion of the collective excitations in the superconductor. If $kv_F \ll \Delta$, the solution of (13.30) is

$$\omega^2 = k^2 u^2, \qquad u^2 = \tfrac{1}{3} v_F{}^2 \tag{13.31}$$

and corresponds to acoustic oscillations. Therefore, we have for the correlation function of the density fluctuations associated with the acoustic branch of the collective excitations

$$\langle \delta n^2 \rangle_{\mathbf{k}\omega} = \frac{2\pi n k^2}{m} \delta(\omega^2 - k^2 u^2), \qquad -2\Delta < \omega < 0. \tag{13.32}$$

In the domain of the continuous spectrum $\omega < -2\Delta$ the correlation function (13.27) already does not have the form of a delta function. However, near the curve $\omega(k)$, which corresponds to the solution of (13.30), there is a sharp maximum, particularly near the boundary of the continuous spectrum. Far from this curve, the correlation function (13.27) is no different than (13.16) in practice. The ratio of the contribution of the collective fluctuations to the total fluctuations is

$$\int_{-2\Delta}^0 \langle \delta n^2 \rangle_{\mathbf{k}\omega} \, d\omega \; \Big| \int_{-\infty}^0 \langle \delta n^2 \rangle_{\mathbf{k}\omega} \, d\omega = \frac{ku}{6\pi\Delta + ku} . \tag{13.33}$$

According to the condition for applicability of the analysis $kv_F \ll \Delta$, this ratio is small.

In the case of charged particles with $k \to 0$ Coulomb interaction plays an essential part. Taking account of Coulomb interaction leads to the following expression for the correlation function:

$$\langle \delta n^2 \rangle_{\mathbf{k}\omega} = \frac{2}{V(k)} \frac{1}{e^{\omega/T} - 1} \operatorname{Im} \frac{\theta\theta^* - (b^* - 1)\,d}{b^* d - \theta\theta^*}, \tag{13.34}$$

where b^* and θ^* differ from (13.28) by the replacement of I by $V(k) = -4\pi e^2 / k^2$.

4. Current Fluctuations in a Superconductor

In an analogous manner it is easy to consider the current density fluctuations of electrons in a superconductor. Defining the current density by the relationship (12.14), and utilizing (13.6), we obtain the following general formula for the spectral distribution of the correlation function of the transverse current fluctuations at arbitrary temperature:

$$\langle \delta j_t^2 \rangle_{\mathbf{k}\omega} = (e^{\omega/T} - 1)^{-1} \frac{4\pi}{V} \sum_{\mathbf{p}} \frac{1}{m^2} \left(p^2 - \frac{(\mathbf{kp})^2}{k^2} \right) \{ (N_{\mathbf{p}} - N_{\mathbf{p}-\mathbf{k}})(u_{\mathbf{p}} u_{\mathbf{p}-\mathbf{k}}$$

$$+ v_{\mathbf{p}} v_{\mathbf{p}-\mathbf{k}})^2 [\delta(\omega - E_{\mathbf{p}-\mathbf{k}} + E_{\mathbf{p}}) - \delta(\omega - E_{\mathbf{p}} + E_{\mathbf{p}-\mathbf{k}})]$$

$$+ (1 - N_{\mathbf{p}} - N_{\mathbf{p}-\mathbf{k}})(u_{\mathbf{p}} v_{\mathbf{p}-\mathbf{k}} - v_{\mathbf{p}} u_{\mathbf{p}-\mathbf{k}})^2$$

$$\times [\delta(\omega - E_{\mathbf{p}} - E_{\mathbf{p}-\mathbf{k}}) - \delta(\omega + E_{\mathbf{p}} + E_{\mathbf{p}-\mathbf{k}})] \}. \tag{13.35}$$

Here

$$(u_{\mathbf{p}} u_{\mathbf{p}-\mathbf{k}} + v_{\mathbf{p}} v_{\mathbf{p}-\mathbf{k}})^2 = \frac{1}{2} \left(1 + \frac{\xi_{\mathbf{p}} \xi_{\mathbf{p}-\mathbf{k}} + \varDelta^2}{E_{\mathbf{p}} E_{\mathbf{p}-\mathbf{k}}} \right),$$

$$(u_{\mathbf{p}} v_{\mathbf{p}-\mathbf{k}} - v_{\mathbf{p}} u_{\mathbf{p}-\mathbf{k}})^2 = \frac{1}{2} \left(1 - \frac{\xi_{\mathbf{p}} \xi_{\mathbf{p}-\mathbf{k}} + \varDelta^2}{E_{\mathbf{p}} E_{\mathbf{p}-\mathbf{k}}} \right). \tag{13.36}$$

If the temperature equals zero $(T = 0)$, formula (13.35) simplifies to

$$\langle \delta j_t^2 \rangle_{\mathbf{k}\omega} = \frac{1}{8\pi^2 m^2} \int d\mathbf{p} \left(p^2 - \frac{(\mathbf{kp})^2}{k^2} \right) \left(1 - \frac{\xi_{\mathbf{p}} \xi_{\mathbf{p}-\mathbf{k}} + \varDelta^2}{E_{\mathbf{p}} E_{\mathbf{p}-\mathbf{k}}} \right)$$

$$\times \delta(\omega + E_{\mathbf{p}} + E_{\mathbf{p}-\mathbf{k}}). \tag{13.37}$$

If $k = 0$, then $\langle \delta j_t^2 \rangle_{0\omega} = 0$. This is related to the fact that the correlations for the parallel and antiparallel states cancel. For large values of k the spectral density decreases more rapidly than k^{-2}. For a quadratic dispersion law in the ranges of values of ω and k satisfying the conditions

$$\omega(\omega + 2\varDelta) < E_F^2,$$

$$p_F - \{p_F^2 - 2m[\omega(\omega + 2\varDelta)]^{1/2}\}^{1/2} < k < p_F + \{p_F^2 - 2m[\omega(\omega + 2\varDelta)]^{1/2}\}^{1/2},$$

we have

$$\langle \delta j_t^2 \rangle_{\mathbf{k}\omega} = \frac{2\varDelta p_F^2}{\pi k} \left\{ \left(1 - \frac{k^2}{4p_F^2} \right) \left[(s + 1)E \left(\frac{s-1}{s+1} \right) - 2K \left(\frac{s-1}{s+1} \right) \right] \right.$$

$$- \frac{\varDelta^2}{E_F^2} \frac{p_F^2}{k^2} \left[(s + 1)(s^2 + 1)E \left(\frac{s-1}{s+1} \right) - (2s^2 + s + 1) \right.$$

$$\left. \left. K \left(\frac{s-1}{s+1} \right) \right] \right\}, \qquad s = \frac{|\omega|}{2\varDelta}. \tag{13.38}$$

The spectral density of the time correlation function in the frequency range $\omega(\omega + 2\Delta) \ll E_F{}^2$ ($|\omega| > 2\Delta$) is also expressed, for $T = 0$, in terms of elliptic integrals

$$\langle \delta j_t{}^2 \rangle_\omega = \frac{8\Delta p_F{}^4}{3\pi^3} \left\{ (s + 1)E \left(\frac{s - 1}{s + 1} \right) - 2K \left(\frac{s - 1}{s + 1} \right) \right\}, \quad s = \frac{|\omega|}{2\Delta}. \quad (13.39)$$

5. Electromagnetic Properties of a Superconductor

The spectral distributions of the correlation functions found in the previous sections may be used to find the dielectric permittivities of a superconductor. Substituting (13.16) and (13.35) into (2.54) and (2.55), we obtain the following formulas for the longitudinal and transverse dielectric permittivities of a superconductor:

$$\epsilon_l(\omega, \mathbf{k}) = 1 - \frac{4\pi e^2}{k^2} \frac{1}{2V} \sum_{\mathbf{p}} \left\{ (N_{\mathbf{p}} - N_{\mathbf{p-k}}) \left(1 + \frac{\xi_{\mathbf{p}} \xi_{\mathbf{p-k}} - \Delta^2}{E_{\mathbf{p}} E_{\mathbf{p-k}}} \right) \right.$$

$$\times \left(\frac{1}{\omega - E_{\mathbf{p-k}} + E_{\mathbf{p}} + io} - \frac{1}{\omega - E_{\mathbf{p}} + E_{\mathbf{p-k}} + io} \right)$$

$$+ (1 - N_{\mathbf{p}} - N_{\mathbf{p-k}}) \left(1 - \frac{\xi_{\mathbf{p}} \xi_{\mathbf{p-k}} - \Delta^2}{E_{\mathbf{p}} E_{\mathbf{p-k}}} \right)$$

$$\left. \times \left(\frac{1}{\omega - E_{\mathbf{p}} - E_{\mathbf{p-k}} + io} - \frac{1}{\omega + E_{\mathbf{p}} + E_{\mathbf{p-k}} + io} \right) \right\}, \quad (13.40)$$

$$\epsilon_t(\omega, k) = 1 - \frac{\Omega^2}{\omega^2} - \frac{2\pi e^2}{\omega^2 m^2} \frac{1}{2V} \sum_{\mathbf{p}} \left(\mathbf{p}^2 - \frac{(\mathbf{kp})^2}{k^2} \right)$$

$$\times \left\{ (N_{\mathbf{p}} - N_{\mathbf{p-k}}) \left(1 + \frac{\xi_{\mathbf{p}} \xi_{\mathbf{p-k}} + \Delta^2}{E_{\mathbf{p}} E_{\mathbf{p-k}}} \right) \right.$$

$$\times \left(\frac{1}{\omega - E_{\mathbf{p-k}} + E_{\mathbf{p}} + io} - \frac{1}{\omega - E_{\mathbf{p}} + E_{\mathbf{p-k}} + io} \right)$$

$$+ (1 - N_{\mathbf{p}} - N_{\mathbf{p-k}}) \left(1 - \frac{\xi_{\mathbf{p}} \xi_{\mathbf{p-k}} + \Delta^2}{E_{\mathbf{p}} E_{\mathbf{p-k}}} \right)$$

$$\left. \times \left(\frac{1}{\omega - E_{\mathbf{p}} - E_{\mathbf{p-k}} + io} - \frac{1}{\omega + E_{\mathbf{p}} + E_{\mathbf{p-k}} + io} \right) \right\}. \quad (13.41)$$

For ω and $kv_F \gg \Delta$ the expressions (13.40) and (13.41) go over into the corresponding expressions for the dielectric permittivities of a normal plasma (12.22) and (12.23). If ω and $kv_F \sim \Delta$, the presence of gaps turns out to be very essential.

It should be expected that ϵ_l and ϵ_t for the superconducting state with $\omega \gg \Delta$ and $kv_F \ll \Delta$ should not differ very radically from the corresponding expressions for the normal plasma since the long-range Coulomb interaction plays the main part in this case. Indeed, the transverse dielectric permittivity of a super-conductor (13.41) turns out to equal $\epsilon_t = 1 - (\Omega^2/\omega^2)$ for $k \to 0$. The longitudinal dielectric permittivity determined by (13.40) does not go over into the corresponding expression for the normal case $\epsilon_l = 1 - (\Omega^2/\omega^2)$ as $k \to 0$. This is related to the fact that collective excitations not taken into account in (13.40) play an essential part in the density fluctuations with small k.

Using (13.34) and (2.54) we find for the longitudinal dielectric permittivity of a superconducting plasma at $T = 0$

$$\epsilon_l(\omega, \mathbf{k}) = 1 - \frac{4\pi e^2}{k^2} \frac{1}{2V} \sum_{\mathbf{p}} \left\{ \left(1 - \frac{\xi_{\mathbf{p}}\xi_{\mathbf{p}-\mathbf{k}} - \Delta^2}{E_{\mathbf{p}}E_{\mathbf{p}-\mathbf{k}}} \right) - \frac{\omega\Delta}{2E_{\mathbf{p}}E_{\mathbf{p}-\mathbf{k}}} \frac{\theta}{d} \right\}$$

$$\times \left(\frac{1}{\omega - E_{\mathbf{p}} - E_{\mathbf{p}-\mathbf{k}} + io} - \frac{1}{\omega + E_{\mathbf{p}} + E_{\mathbf{p}-\mathbf{k}} + io} \right). \quad (13.42)$$

This expression differs from (13.40) at $T = 0$ by the additional member in the braces, which is very essential near the curve $\omega(k)$ corresponding to the acoustic oscillations. Far from this curve the expressions (13.42) and (13.40) practically agree for $T = 0$. If $|\omega| \gg \Delta$ and $kv_F \ll \Delta$, then we obtain $\epsilon_l = 1 - (\Omega^2/\omega^2)$ from (13.42). Therefore, the condition $\epsilon_l(\omega, \mathbf{k}) = 0$ leads to the custom-ary plasma oscillations for small k. The correlation function of the density fluctuations of a superconducting plasma with $kv_F \ll \Delta$ agrees with the corresponding expression for a normal plasma near the plasma oscillation frequencies

$$\langle \delta n^2 \rangle_{\mathbf{k}\omega} = \frac{2\pi n_0 k^2}{m} \delta(\omega^2 - \Omega^2). \quad (13.43)$$

The behavior of a superconductor in an external transverse electromagnetic field is determined by the dielectric permittivity

$\epsilon_t(\omega, \mathbf{k})$. In particular, the conductivity of a superconductor $\sigma(\omega, \mathbf{k})$ is connected with $\epsilon_t(\omega, \mathbf{k})$ by means of the relationship

$$\sigma(\omega, \mathbf{k}) = \frac{\omega}{4\pi i}\{\epsilon_t(\omega, \mathbf{k}) - 1\}. \tag{13.44}$$

Using (13.41), it is easy to see that the static conductivity of a superconductor becomes infinite at temperatures below some critical temperature T_0. In fact, limiting ourselves to the case $k \to 0$, we find in the low frequency range

$$\text{Re } \sigma(\omega, o) = \frac{1}{4}\left\{\Omega^2 + \frac{4\pi e^2}{V}\sum_{\mathbf{p}} \frac{p^2}{m^2}\frac{\partial N_{\mathbf{p}}}{\partial E_{\mathbf{p}}}\right\}\delta(\omega)$$

$$+ \text{ a finite component as } \omega \to 0. \tag{13.45}$$

The brace in front of the delta function is a function which decreases monotonely as the temperature increases, and vanishes at $T = T_0$. Hence, superconductivity occurs only at temperatures less than the critical temperature T_0, determined from the condition

$$\Omega^2 + \frac{4\pi e^2}{V}\sum_{\mathbf{p}} \frac{p^2}{m^2}\frac{\partial N_{\mathbf{p}}}{\partial E_{\mathbf{p}}} = 0. \tag{13.46}$$

In the limiting case of small depth of penetration of the field (the depth of penetration is small compared with the coherence length v_F/Δ) the electromagnetic properties of the superconductor are characterized by the conductivity $\sigma(\omega)$, determined by the integral of (13.44) with respect to k. Utilizing (13.39), we obtain for the real and imaginary parts of the conductivity $\sigma(\omega)$ at $T = 0$ (5):

$$\frac{\sigma'}{\sigma_n} = \frac{1}{s}\left\{(s+1)E\left(\frac{s-1}{s+1}\right) - 2K\left(\frac{s-1}{s-1}\right)\right\}, \quad s = \frac{|\omega|}{2\Delta} > 1, \tag{13.47}$$

$$\frac{\sigma''}{\sigma_n} = \frac{1}{2s}\left\{(s+1)E\left(\frac{2\sqrt{s}}{s+1}\right) - (s-1)K\left(\frac{2\sqrt{s}}{s+1}\right)\right\}, \quad s = \frac{|\omega|}{2\Delta},$$

$$\tag{13.48}$$

where σ_n is the conductivity in the normal state.

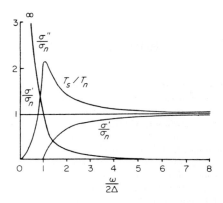

FIG. 27.

Dependences of the real and imaginary parts of the conductivity $\sigma(\omega)$ on the nondimensional frequency $\omega/2\Delta$ are presented in Fig. 27. Also presented in this same figure is the frequency dependence of the coefficient of radiation passage through a thin superconducting film (9)

$$\frac{T_s}{T_n} = \left\{ \left[T_n^{1/2} + (1 - T_n^{1/2}) \frac{\sigma'}{\sigma_n} \right]^2 + \left[(1 - T_n^{1/2})^2 \frac{\sigma''}{\sigma_n} \right]^2 \right\}^{-1}. \tag{13.49}$$

(T_n is the coefficient of passage through the film in the normal state.) The transmission coefficient T_n is characterized by a sharp maximum at a frequency double the gap width ($\omega \simeq 2\Delta$).

REFERENCES

1. J. Bardeen, L. Cooper, and J. Schrieffer, *Phys. Rev.* **108**, 1175 (1957).
2. N. N. Bogoliubov, *ZETF* **34**, 58 (1958) *(Sov. Phys. JETP)*.
3. N. N. Bogoliubov, V. V. Tolmachev, and D. V. Shirkov, "New Methods in Superconductivity Theory." AN SSSR Press, Moscow, 1958 (CFSTI).
4. A. A. Abrikosov, L. P. Gor'kov, and I. M. Khalatnikov, *ZETF* **35**, 265 (1958); **37**, 187 (1959) *(Sov. Phys. JETP)*.
5. D. Mattis and J. Bardeen, *Phys. Rev.* **111**, 412 (1958).
6. G. Rickaysen, *Phys. Rev.* **115**, 795 (1959).
7. L. P. Kadanoff and P. C. Martin, *Phys. Rev.* **124**, 670 (1961).
8. A. G. Sitenko and I. V. Simenog, *Ukr. Fiz. Zh.* **8**, 537 (1963).
9. J. Bardeen and J. Schrieffer, "New Methods in the Study of Superconductivity." Fizmatgiz, Moscow, 1962.

BIBLIOGRAPHY

Translator's Note: Insofar as has been possible, in this Bibliography, I have tried to check the references, and to determine whether translations of the Russian articles exist, and if so, to indicate where they may be obtained.

Prior to 1954, the translation of foreign technical articles was on a hit-or-miss basis by both industrial and educational organizations. Some of these translations were collected, catalogued, and made available by the Special Libraries Association (SLA) from its Chicago office. There were also sporadic translation efforts funded by the National Science Foundation (NSF) and the Brookhaven National Laboratory (BNL), and these translations were, I think, catalogued and made available by the Library of Congress (LC), which also tried to compile a bibliography of technical and scientific translations available from any and every source, including the commercial translation organizations.

In 1955, the American Institute of Physics (AIP) was awarded a grant to commence the translation of the Zhurnal Eksperimental'noi i Teoreticheskoi Fiziki (ZETF) as Soviet Physics JETP, and under the supervision of Professor R. Beyer, these translations started to appear on a first-translated, first-published basis. Later, this disorder was regularized under J. G. Adashko's direction, and the JETP now appear in cover-to-cover translation in the same order that the Russian articles appear in the ZETF.

Subsequently, the AIP added other Soviet Physics publications [translated by Consultants Bureau Enterprises (CB)] to its ever-increasing list, namely: Doklady, Technical Physics (ZTF),* Astronomy (AZ), Solid State (FTT), etc. Other Russian journals were also being translated cover-to-cover by other professional groups (FMM, Physics of Metals and Metallography) at the same time, and wherever I have been able, I indicate the volume and page and year of the English equivalent for the referenced articles. In all cases, where translations exist, I indicate the source from which they may be obtained.

After the AIP had commenced the translation and publication of the Uspekhi, the NSF and AEC decided jointly to have issues translated which had been published prior to the beginning of the AIP series. These volumes are now available through the Clearinghouse for Federal Scientific and Technical Information (CFSTI), which also makes available translations of single articles and complete Soviet journals executed by the United States Joint Publication Research Service (JPRS) at the request of divers government agencies, as well as the previous holdings of Morris D. Friedman, Inc. (MDF), no longer extant. The CFSTI also took over the task of cataloging all available translations and of publishing this listing in its biweekly Technical Translations magazine.

Since adequate facilities, and time, for a comprehensive search as to the existence of translated articles from years prior to the commencement of cover-to-cover

---

* ZTF—Zhurnal Tekhnicheskoi Fiziki.

translations were not at hand, such a search was not conducted. In any case, it would be impossible (unless all the early papers listed had already been translated prior to this writing) to include a comprehensive list, since articles are constantly being translated, and some of the untranslated ones may very well be translated in the near future.

I have used the Library of Congress transliteration scheme throughout. At present, and I guess always, a number of transliterations schemes exist and it is impossible to know which to select. Indeed, at the beginning of its translation program the AIP also used the LC transliteration scheme, but it has now accepted the British scheme, which differs from LC primarily by its utilization of y instead of i in the diphthong sounds "ia", "iu", etc.

Below is a listing of some of the translation organizations mentioned in the Bibliography, and their current addresses.

CFSTI Clearinghouse for Federal Scientific and Technical Information
U.S. Department of Commerce
5285 Port Royal Road
Springfield, Virginia 22151

CB Consultants Bureau Enterprises
227 West 17th Street
New York, N.Y. 10011

GB Gordon & Breach
150 Fifth Avenue
New York, N.Y. 10011

IP Wiley (Interscience Publishers)
605 Third Avenue
New York, N.Y. 10016

PP Pergamon Press
122 East 55th Street
New York, N.Y.

SLA Special Libraries Association
John Crerar Library
35 West 33rd Street
Chicago, Illinois 60616

LC Photoduplication Service
Library of Congress
Washington, D.C. 20025

ATS Associated Technical Services
Post Office Box 271
East Orange, New Jersey 07017

AIP American Institute of Physics
335 East 45th Street
New York, New York 10017

1. Agranovich, V. M., Pafomov, V. E., and Rukhadze, A. A., On Cerenkov radiation of an electron moving in a medium with spatial dispersion, *ZETF* **36**, 238 (1959) [*JETP* **9**, 160 (1959)].

2. Agranovich, V. M. and Rukhadze, A. A., On electromagnetic wave propagation in a medium taking account of spatial dispersion, *ZETF* **35**, 982 (1958) [*JETP* **8**, 685 (1959)].

3. Agranovich, V. M. and Rukhadze, A. A., Energy losses of electrons in a medium with spatial dispersion, *ZETF* **35**, 1171 (1958) [*JETP* **8**, 819 (1959)].

4. Alekseev, A. I., Application of quantum field theory methods in statistical physics, *UFN* **73**, 40 (1961) [*Uspekhi* **4**, 23 (1961)].

5. Aleksin, V. F. and Stepanov, K. N., On the theory of electromagnetic fluctuations in a plasma, *Izv. Vuz, Radiofizika* **5**, 61 (1962) [CFSTI].

6. Aleksin, V. F. and Iashin, V. I., Investigation of plasma stability by using a generalized energy principle, *ZETF* **39**, 822 (1960) [*JETP* **12**, 572 (1961)].

7. Aleksin, V. F. and Iashin, V. I., On the stability of a plasma with anisotropic particle velocity distribution and arbitrarily distributed current, *ZETF* **40**, 1115 (1961) [*JETP* **13**, 787 (1961)].

8. Al'pert, Ia.L., Ginzburg, V.L., and Feinberg, E.L., "Radiowave Propagation," Gostekhizdat, Moscow, 1953 (Selected chapters translated by MDF and available from CFSTI).

9. Aliamovskii, V. N., On distribution functions of an electric microfield in an ionized gas, *ZETF* **42**, 1536 (1962) [*JETP* **15**, 1067 (1962)].

10. Andronov, A. A., On the question of the damping and growth of plasma waves, *Izv. Vuz, Radiofizika* **4**, 861 (1961) [CFSTI].

11. Artsimovich, L.A., "Controlled Thermonuclear Fusion," Fizmatgiz, Moscow, 1961 (IP, 1961).

12. Askar'ian, G. A., Acceleration of ionized gas clouds with a frozen magnetic field dissipating the electron flux, *ZETF* **41**, 1632 (1961) [*JETP* **14**, 1159 (1962)].

13. Akhiezer, A. I., Aleksin, V. F., Bar'iakhtar, V. G., and Peletminskii, S. V., Influence of radiation effects on relaxation of plasma electrons and electrical conductivity in a strong magnetic field, *ZETF* **42**, 552 (1962) [*JETP* **15**, 386 (1962)].

14. Akhiezer, A. I., Akhiezer, I. A., and Sitenko, A. G., On the theory of fluctuations in a plasma, *ZETF* **41**, 644 (1962) [*JETP* **14**, 462 (1962)].

15. Akhiezer, A. I., Bar'iakhtar, V. G., and Peletminskii, S. V., On the influence of radiation processes on transport phenomena in a plasma in a strong magnetic field, *ZETF* **43**, 1743 (1962) [*JETP* **16**, 1231 (1963)].

16. Akhiezer, A. I., Kitsenko, A. B., and Stepanov, K. N., On the interaction of charged particle streams with low-frequency plasma oscillations, *ZETF* **40**, 1866 (1961) [*JETP* **13**, 1311 (1961)].

17. Akhiezer, A. I. and Liubarskii, G. Ia., On the stability of distribution functions of an electron plasma, *Uchen. Zapisk. Khar'k. Gos. Univ.* **64**, 13 (1955).

18. Akhiezer, A. I., Liubarskii, G. Ia., and Polovin, R. V., On shockwave stability in magnetohydrodynamics, *ZETF* **35**, 731 (1958) [*JETP* **8**, 507 (1959)].

19. Akhiezer, A. I., Liubarskii, G. Ia., and Polovin, R. V., Simple waves in magnetohydrodynamics, *Ukr. Fiz. Zh.* **3**, 433 (1958) [MDF—CFSTI].

20. Akhiezer, A. I., Liubarskii, G. Ia., and Polovin, R. V., On stability conditions of electron distribution functions in a plasma, *ZETF* **40**, 963 (1961) [*JETP* **13**, 673 (1961)].

21. Akhiezer, A. I., Liubarskii, G. Ia., and Fainberg, Ia. B., On the nonlinear theory of oscillations in plasma, *Uchen. Zapisk. Khar'k. Gos. Univ.* **64**, 73 (1955).

22. Akhiezer, A. I., and Pargamanik, L. E., Free oscillations of an electron plasma in a magnetic field, *Uchen. Zapisk. Khar'k. Gos. Univ.* **27**, 75 (1948) [CFSTI].

23. Akhiezer, A. I. and Polovin, R. V., On plasma oscillations in crossed electric and magnetic fields, *ZTF* **22**, 1794 (1952).

24. Akhiezer, A. I. and Polovin, R. V., On relativistic plasma oscillations, *DAN SSSR* **102**, 919 (1955).

25. Akhiezer, A. I. and Polovin, R. V., On the theory of wave motions of relativistic plasma, *ZETF* **30**, 915 (1956) [*JETP* **3**, 696 (1956)].

26. Akhiezer, A. I., Prokhoda, I. G., and Sitenko, A. G., Electromagnetic wave scattering in plasmas, *ZETF* **33**, 750 (1957) [*JETP* **6**, 576 (1958)].

27. Akhiezer, A. I. and Sitenko, A. G., On the passage of charged particles through a plasma, *ZETF* **23**, 161 (1952).

28. Akhiezer, A. I. and Sitenko, A. G., On the theory of resonant cavity excitation by moving charged particles, *ZTF* **23**, 1217 (1953).

29. Akhiezer, A. I. and Sitenko, A. G., On oscillations of an electron plasma in an external electric field, *ZETF* **30**, 216 (1956) [*JETP* **3**, 140 (1956)].

30. Akhiezer, A. I. and Sitenko, A. G., On the theory of hydromagnetic wave excitation, *ZETF* **35**, 116 (1958) [*JETP* **8**, 82 (1959)].

31. Akhiezer, A. I. and Sitenko, A. G., On the theory of hydromagnetic wave excitation: "Questions of Magnetohydrodynamics and Plasma Dynamics," p. 137. AN LatvSSR Press, Riga, 1959 [CFSTI].

32. Akhiezer, A. I. and Fainberg, Ia. B., On the interaction between a beam of charged particles and an electron plasma, *DAN SSSR* **69**, 555 (1949) [CFSTI].

33. Akhiezer, A. I. and Fainberg, Ia. B., On high-frequency oscillations of an electron plasma, *ZETF* **21**, 1262 (1951).

34. Akhiezer, A. I. and Fainberg, Ia. B., Slow electromagnetic waves, *UFN* **44**, 321 (1951).

35. Akhiezer, A. I., Fainberg, Ia. B., Sitenko, A. G., Stepanov, K. N., Kurilko, V. I., Gorbatenko, M. F., and Kirochkin, Iu. A., On the theory of high-frequency plasma oscillations, "Proceedings of the Second International Conference on the Peaceful Uses of Atomic Energy," Geneva, 1958 [CB].

36. Akhiezer, I. A., On the theory of charged particle interaction with a plasma in a magnetic field, *ZETF* **40**, 954 (1961) [*JETP* **13**, 667 (1961)].

37. Akhiezer, I. A., On the theory of the electromagnetic properties of a system of many charged particles, *UFZh* **6**, 435 (1961).

38. Akhiezer, I. A., On the development of fluctuations in a plasma with an unstable distribution function, *ZETF* **42**, 584 (1962) [*JETP* **15**, 406 (1962)].

39. Akhiezer, I. A. and Peletminskii, S. V., On application of quantum field

theory methods to investigation of thermodynamic properties of an electron and photon gas, *ZETF* **38**, 1829 (1960) [*JETP* **11**, 1316 (1960)].

40. Akhiezer, I. A. and Peletminskii, S. V., On the theory of the magnetic properties of a non-ideal Fermi gas at low temperatures, *ZETF* **39**, 1308 (1960) [*JETP* **12**, 913 (1961)].

41. Akhiezer, I. A. and Polovin, R. V., On the theory of relativistic magneto-hydrodynamic waves, *ZETF* **36**, 1845 (1959) [*JETP* **9**, 1316 (1959)].

42. Akhiezer, I. A., Polovin, R. V., and Tsintsadze, N. L., Simple waves in the Chu, Goldberger, and Low approximation, *ZETF* **37**, 756 (1959) [*JETP* **10**, 539 (1960)].

43. Barabanenkov, Iu. N., On the solution of the kinetic equation for a plasma in an alternating magnetic field, *ZETF* **37**, 427 (1959) [*JETP* **10**, 305 (1960)].

44. Barabanenkov, Iu. N., On the possibility of charge self-oscillation in crossed fields, *ZETF* **38**, 263 (1960) [*JETP* **11**, 190 (1960)].

45. Barsukov, K. A., On the Doppler effect in an anisotropic and gyrotropic medium, *ZETF* **36**, 1485 (1959) [*JETP* **9**, 1057 (1959)].

46. Bass, F. G. and Blank, A. Ia., On the theory of wave transformation and scattering by fluctuations in a plasma, *ZETF* **43**, 1479 (1962) [*JETP* **16**, 1045 (1963)].

47. Bass, F. G. and Kaganov, M. I., Correlation relationships for random currents and fields at low temperatures, *ZETF* **34**, 1154 (1958) [*JETP* **7**, 799 (1958)].

48. Beliaev, S. T., Kinetic equation for rarefied gases in strong fields; "Plasma Physics and the Problem of Controlled Thermonuclear Reactions," Vol. 3, p. 50. AN SSSR Press, Moscow, 1958 [PP].

49. Beliaev, S. T. and Budker, G. I., Relativistic kinetic equation, *DAN SSSR* **107**, 807 (1956) [*Sov. Phys. Dokl.* **1**, 218 (1956)].

50. Beliaev, S. T. and Budker, G. I., Relativistic plasma in alternating fields; "Plasma Physics and the Problem of Controlled Thermonuclear Reactions," Vol. 2, p. 283. AN SSSR Press, Moscow, 1958 [PP].

51. Bogdankevich, L. S., Radiation of a current ring moving uniformly in a plasma in a magnetic field, *ZETF* **36**, 835 (1959) [*JETP* **9**, 589 (1959)].

52. Bogoliubov, N. N. and Zubarev, D. N., Method of asymptotic approximation for systems with rotating phase and its application to charged particle motion in a magnetic field, *UMZh* **7**, 5 (1955) [*LC*; *SLA*].

53. Bogoliubov, N. N. and Tiablikov, S. V., Retarded and advanced Green's functions in statistical physics, *DAN SSSR* **126**, 53 (1959) [*Sov. Phys. Dokl.* **4**, 589 (1959)].

54. Bolotovskii, B. M., Theory of the Vavilov-Cerenkov effect, *UFN* **62**, 201 (1957) [CFSTI].

55. Bolotovskii, B. M. and Rukhadze, A. A., Field of charged particles in a moving medium, *ZETF* **37**, 1346 (1959) [*JETP* **10**, 958 (1960)].

56. Bonch-Bruevich, V. L., On the theory of electron-gas interaction with crystal-lattice oscillations, *ZETF* **30**, 342 (1956) [*JETP* **3**, 278 (1956)].

57. Bonch-Bruevich, V. L., Remark on the theory of the electron plasma in semiconductors, *ZETF* **32**, 1092 (1957) [*JETP* **5**, 894 (1957)].

58. Bonch-Bruevich, V. L., On application of the Green's function method to

the multielectron problem in solid state, *FMM* **6**, 590 (1958) [*Physics of Metals and Metallography* **6**(3) (1958)].

59. Bonch-Bruevich, V. L. and Mironov, A. G., Theory of an electron plasma in a magnetic field, *FTT* **2**, 489 (1960) [*Sov. Phys. Solid State* **2**, 454 (1960)].

60. Bonch-Bruevich, V. L. and Tiablikov, S. V., "Green's Function Method in Statistical Mechanics," Moscow, Fizmatgiz, 1961 [Wiley (Interscience), 1962].

61. Braginskii, S. I., On the theory of charged particle motion in a strong magnetic field, *UMZh* **8**, 119 (1956).

62. Braginskii, S. I., On kinds of plasma oscillations in a magnetic field, *DAN SSSR* **115**, 475 (1957) [*Sov. Phys. Dokl.* **2**, 345 (1957)].

63. Braginskii, S. I., Transport phenomena in a fully ionized two-temperature plasma, *ZETF* **33**, 459 (1957) [*JETP* **6**, 358 (1958)].

64. Braginskii, S. I., On the behavior of a fully ionized plasma in a strong magnetic field, *ZETF* **33**, 645 (1957) [*JETP* **6**, 494 (1958)].

65. Braginskii, S. I., Particle and heat fluxes across a strong magnetic field in a fully ionized two-temperature plasma; "Plasma Physics and the Problem of Controlled Thermonuclear Reactions," Vol. 1, Moscow, AN SSSR Press, 1958 [PP].

66. Braginskii, S. I., On the magnetohydrodynamics of weakly conductive fluids, *ZETF* **37**, 1417 (1959) [*JETP* **10**, 1005 (1960)].

67. Braginskii, S. I. and Kazantsev, A. P., Magnetohydrodynamic waves in a rarefied plasma; "Plasma Physics and the Problem of Controlled Thermonuclear Reactions," Vol. 4, p. 24. Moscow, AN SSSR Press, 1958 [PP].

68. Budker, G. I., Relativistic stabilized electron beams, *AE* **5**, 9 (1956) [CB].

69. Budker, G. I. and Beliaev, S. T., Kinetic equation for an electron gas with rare collisions; "Plasma Physics and the Problem of Controlled Thermonuclear Reactions," Vol. 2, Moscow, AN SSSR Press, 1958 [PP].

70. Bunkin, F. V., Thermal radiation of an anisotropic medium, *ZETF* **32**, 811 (1957) [*JETP* **5**, 665 (1957)].

71. Bunkin, F. V., On radiation in anisotropic media, *ZETF* **32**, 338 (1957) [*JETP* **5**, 277 (1957)].

72. Bunkin, F. V., On the theory of electromagnetic fluctuations in a nonequilibrium plasma, *ZETF* **41**, 288 (1961) [*JETP* **14**, 206 (1962)].

73. Bunkin, F. V., On the theory of electromagnetic fluctuations in a nonstationary plasma, *ZETF* **41**, 1859 (1961) [*JETP* **14**, 1322 (1962)].

74. Vedenov, A. A., On some solutions of the plasma hydrodynamic equations, *ZETF* **33**, 1509 (1957) [*JETP* **6**, 1165 (1958)].

75. Vedenov, A. A., Thermodynamic properties of a degenerate plasma, *ZETF* **36**, 641 (1959) [*JETP* **9**, 446 (1959)].

76. Vedenov, A. A., Velikhov, E. P., and Sagdeev, R. Z., Plasma stability, *UFN* **73**, 701 (1961) [*Sov. Phys. Uspekhi* **4**, 332 (1961)].

77. Vedenov, A. A. and Larkin, A. I., Plasma equation of state, *ZETF* **36**, 1133 (1959) [*JETP* **9**, 806 (1959)].

78. Vedenov, A. A. and Rudakov, L. N., On charged particle motion in rapidly alternating electromagnetic fields; "Plasma Physics and the Problem of Controlled Thermonuclear Reactions," Vol. 4, Moscow, AN SSSR Press, 1958 [PP].

79. Vedenov, A. A. and Sagdeev, R. Z., On some properties of a plasma with an anisotropic ion-velocity distribution in a magnetic field; "Plasma Physics and the Problem of Controlled Thermonuclear Reactions," Vol. 3, p. 278. Moscow, AN SSSR Press, 1958 [PP].

80. Vlasov, A. A., On the vibrational properties of an electron gas, *ZETF* **8**, 291 (1938).

81. Vlasov, A. A., Theory of the vibrational properties of an electron gas and its application, *Uch. Zap. Moscow Univ.* **75**, Vyp. 2 (1945).

82. Vlasov, A. A., "Theory of Many Particles," Moscow, Gostekhizdat, 1945 [GB, 1961].

83. Volkov, T. F., Influence of a high-frequency electromagnetic field on plasma oscillations; "Plasma Physics and the Problem of Controlled Thermonuclear Reactions," Vol. 4, Moscow, AN SSSR Press, 1958 [PP].

84. Volkov, T. F., On ion oscillations in a plasma, *ZETF* **37**, 422 (1959) [*JETP* **10**, 302 (1960)].

85. Voloshinskii, A. and Kobelev, L. Ia., On dispersion relations for an electron plasma, *FMM* **6**, 356 (1958) [*FMM* **6**(2), 180 (1958)].

86. Galaiko, V. P. and Pargamanik, L. E., On correlation functions for a system of identical charged particles, *DAN SSSR* **123**, 999 (1958) [*Sov. Phys. Dokl.* **3**, 1225 (1958)].

87. Galitskii, V. M., Energy spectrum of a nonideal Fermi gas, *ZETF* **34**, 151 (1958) [*JETP* **7**, 104 (1958)].

88. Galitskii, V. M. and Migdal, A. B., Application of quantum field theory methods to the many-body problem, *ZETF* **34**, 139 (1958) [*JETP* **7**, 96 (1958)].

89. Galitskii, V. M. and Migdal, A. B., Dielectric constant of a high-temperature magnetized plasma and estimation of the radiant heat conduction; "Plasma Physics and the Problem of Controlled Thermonuclear Reactions," Vol. 1, Moscow, AN SSSR Press, 1958 [PP].

90. Gertsenshtein, M. E., On longitudinal waves in an ionized medium (plasma), *ZETF* **22**, 303 (1952).

91. Gertsenshtein, M. E., Self-excitation of oscillations in the positive column of a gas discharge, *ZETF* **23**, 669 (1952).

92. Gertsenshtein, M. E., Energy relationships in spatially dispersing media, *ZETF* **26**, 680 (1954) [CFSTI].

93. Gertsenshtein, M. E., Dielectric permittivity of a plasma in a stationary magnetic field, *ZETF* **27**, 180 (1954) [SLA].

94. Gertsenshtein, M. E., On the theory of the shot effect, *ZTF* **25**, 827 (1955).

95. Gertsenshtein, M. E., Correlation functions in an electron gas, *ZTF* **25**, 834 (1955).

96. Gertsenshtein, M. E. and Pustovoit, V. I., On high-frequency conductivity of a plasma in the presence of direct current, *ZETF* **43**, 536 (1962) [*JETP* **16**, 383 (1963)].

97. Gershman, B. N., Kinetic theory of magnetohydrodynamic waves, *ZETF* **24**, 453 (1953) [SLA].

98. Gershman, B. N., On electromagnetic wave propagation in a plasma in a

permanent magnetic field, taking account of thermal electron motion, *ZETF* **24**, 659 (1953) [SLA, LC].

99. Gershman, B. N., On normal waves in a homogeneous plasma in a magnetic field; "Collection in Memory of A. A. Andronov," Moscow, AN SSSR Press, 1955 [MDF, CFSTI].

100. Gershman, B. N., Remarks on waves in a homogeneous magnetoactive plasma, *ZETF* **31**, 707 (1956) [*JETP* **4**, 582 (1957)].

101. Gershman, B. N., On the kinetic theory of magnetohydrodynamic wave propagation in a plasma, *Izv. Vuz, Radiofizika* **1**(4), 3 (1958) [CFSTI].

102. Gershman, B. N., On longitudinal waves in a nonisothermal plasma, *Izv. Vuz, Radiofizika* **2**, 654 (1959) [CFSTI].

103. Gershman, B. N., On nonresonant absorption of electromagnetic waves in a magnetoactive plasma, *ZETF* **37**, 695 (1959) [*JETP* **10**, 497 (1960)].

104. Gershman, B. N., On gyroresonant absorptions of electromagnetic waves in plasma, *ZETF* **38**, 912 (1960) [*JETP* **11**, 657 (1960)].

105. Gershman, B. N., On the group velocity of plasma waves in the presence of a magnetic field, *Izv. Vuz, Radiofizika* **3**, 146 (1960) [CFSTI].

106. Gershman, B. N., On the question of electromagnetic wave propagation in a slightly relativistic magnetoactive plasma, *Izv. Vuz, Radiofizika* **3**, 534 (1960) [CFSTI].

107. Gershman, B. N., Ginzburg, V. L., and Denisov, N. G., Electromagnetic wave propagation in a plasma, *UFN* **61**, 561 (1957) [CFSTI].

108. Gershman, B. N. and Kovner, M. S., On singularities of quasi-transverse magnetohydrodynamic wave propagation in a magnetoactive plasma, *Izv. Vuz, Radiofizika* **1**, 19 (1958) [CFSTI].

109. Gershman, B. N. and Kovner, M. S., On some wave propagation peculiarities in a plasma associated with taking account of collisions, *Izv. Vuz, Radiofizika* **2**, 28 (1959) [CFSTI].

110. Getmantsev, G. G., On electromagnetic wave growth in mutually penetrating unboundedly moving media, *ZETF* **37**, 843 (1959) [*JETP* **10**, 600 (1960)].

111. Getmantsev, G. G. and Rapoport, V. O., On electromagnetic wave growth in a plasma moving in a dielectric without dispersion, in the presence of a permanent magnetic field, *ZETF* **38**, 1205 (1960) [*JETP* **11**, 871 (1960)].

112. Ginzburg, V. L., On the electrodynamics of an anisotropic medium, *ZETF* **10**, 601 (1940).

113. Ginzburg, V. L., "Theory of Radiowave Propagation in the Ionosphere," Gostekhizdat, Moscow, 1949.

114. Ginzburg, V. L., On magnetohydrodynamic waves in a gas, *ZETF* **21**, 788 (1951).

115. Ginzburg, V. L., Some questions of the theory of electrical fluctuations, *UFN* **46**, 348 (1952) [CFSTI].

116. Ginzburg, V. L., On electromagnetic waves in isotropic and crystalline media taking account of spatial dispersion of the dielectric permittivity, *ZETF* **34**, 1593 (1958) [*JETP* **7**, 1096 (1958)].

117. Ginzburg, V. L., Some questions of the theory of radiation for faster-than-light motions in a medium, *UFN* **69**, 537 (1959) [*Uspekhi* **2**, 874 (1960)].

118. Ginzburg, V. L., "Electromagnetic Wave Propagation in a Plasma," Fizmatgiz, Moscow, 1960 [GB, 1961; PP, 1964].

119. Ginzburg, V. L. and Gurevich, A. V., Nonlinear phenomena in a plasma in an alternating electromagnetic field, *UFN* **70**, 201 (1960) [*Uspekhi* **3**, 115 (1960–61)].

120. Ginzburg, V. L. and Zhelezniakov, V. V., On possible mechanisms of sporadic solar radio emission. Emission in an isotropic plasma, *Astron. Zh.* **35**, 694 (1958) [*Astronomy* **2**, 653 (1958)].

121. Ginzburg, V. L. and Zhelezniakov, V. V., On electromagnetic wave absorption and emission by a magnetoactive plasma, *Izv. Vuz, Radiofizika* **1**(2), 59 (1958) [CFSTI].

122. Ginzburg, V. L. and Zhelezniakov, V. V., On mechanisms of sporadic solar emission, *Izv. Vuz, Radiofizika* **2**, 9 (1959) [CFSTI].

123. Ginzburg, V. L. and Eidman, V. Ia., On the reactive force of radiation during charge motion in a medium, *ZETF* **36**, 1823 (1959) [*JETP* **9**, 1300 (1959)].

124. Ginzburg, V. L. and Eidman, V. Ia., On some peculiarities of electromagnetic wave radiation by particles moving at a faster-than-light velocity, *Izv. Vuz, Radiofizika* **2**, 331 (1961) [CFSTI].

125. Gintsburg, M. A., On the anomalous Doppler effect in a plasma, *ZETF* **41**, 752 (1961) [*JETP* **14**, 542 (1962)].

126. Gol'dman, I. I., Oscillations of an electron gas with a Fermi distribution function in the degenerate state, *ZETF* **17**, 681 (1947).

127. Gorbatenko, M. F., Electron beam interaction with a plasma in a magnetic field; "Plasma Physics and the Problem of Controlled Thermonuclear Fusion," AN USSR Press, Kiev, 1962.

128. Gordeev, G. V., Plasma and striata oscillations, *ZETF* **22**, 230 (1952) [LC, SLA].

129. Gordeev, G. V., Plasma oscillations in a magnetic field, *ZETF* **23**, 660 (1952) [LC, SLA].

130. Gordeev, G. V., Transverse oscillations of an electron plasma in a permanent magnetic field, *ZETF* **24**, 445 (1953) [LC, SLA].

131. Gordeev, G. V., Low-frequency plasma oscillations, *ZETF* **27**, 18 (1954) [LC, SLA, CB, ATS].

132. Gordeev, G. V., Excitation of plasma oscillations, *ZETF* **27**, 24 (1954) [LC, SLA, ATS].

133. Gurevich, A. V., On the influence of radiowaves on plasma properties, *ZETF* **30**, 1112 (1956) [*JETP* **3**, 895 (1957)].

134. Gurevich, A. V., Simplification of the equations for the electron distribution function in a plasma, *ZETF* **32**, 1237 (1957) [*JETP* **5**, 1006 (1957)].

135. Gurevich, A. V., On the electron temperature in a plasma in an alternating electric field, *ZETF* **35**, 392 (1958) [*JETP* **8**, 271 (1959)].

136. Gurevich, A. V., On some peculiarities of ohmic heating of an electron gas in a plasma, *ZETF* **38**, 116 (1960) [*JETP* **11**, 85 (1960)].

137. Gurevich, A. V., On the question of the quantity of accelerating particles in an ionized gas in the presence of different acceleration mechanisms, *ZETF* **38**, 1596 (1960) [*JETP* **11**, 1150 (1960)].

138. Gurevich, A. V., On the theory of the effect of runaway electrons, *ZETF* **39**, 1296 (1960) [*JETP* **12**, 904 (1961)].
139. Gurevich, A. V., On peculiarities in the behavior of multicharge ions in a plasma, *ZETF* **40**, 1825 (1961) [*JETP* **13**, 1282 (1961)].
140. Gurevich, V. L. and Firsov, Iu. A., Theory of plasma diffusion in a magnetic field, *ZETF* **41**, 1151 (1961) [*JETP* **14**, 822 (1962)].
141. Davydov, B. I., On the theory of electron motion in gases and in semiconductors, *ZETF* **7**, 1069 (1937).
142. Davydov, B. I., On the influence of plasma oscillations on its electrical and heat conductivities; "Plasma Physics and the Problem of Controlled Thermonuclear Reactions," Vol. 1, Moscow, AN SSSR Press, 1958 [PP].
143. Denisov, N. G., On extraordinary and ordinary wave interaction in the ionosphere and the effect of reflected signal multiplication, *ZETF* **29**, 380 (1955) [*JETP* **2**, 342 (1956)].
144. Denisov, N. G., On the question of electromagnetic wave absorption in resonance domains of an inhomogeneous plasma, *ZETF* **34**, 528 (1958) [*JETP* **7**, 364 (1958)].
145. Dnestrovskii, Iu. N. and Kostomarov, D. P., Electromagnetic waves in a half-space filled with a plasma, *ZETF* **39**, 845 (1960) [*JETP* **12**, 587 (1961)].
146. Dnestrovskii, Iu. N. and Kostomarov, D. P., On the dispersion equation for an ordinary wave being propagated in a plasma transverse to an external magnetic field, *ZETF* **40**, 1404 (1961) [*JETP* **13**, 986 (1961)].
147. Dnestrovskii, Iu. N. and Kostomarov, D. P., On the dispersion equation for an extraordinary wave being propagated in a plasma transverse to the external magnetic field, *ZETF* **41**, 1527 (1961) [*JETP* **14**, 1089 (1962)].
148. Dokuchaev, V. P., On growth of magnetohydrodynamic waves in a plasma moving through an ionized gas, *ZETF* **39**, 413 (1960) [*JETP* **12**, 292 (1961)].
149. Dolgopolov, V. V. and Stepanov, K. N., On magnetohydrodynamic wave attenuation in a rarefied plasma, *Ukr. Fiz. Zh.* **5**, 59 (1960).
150. Eleonskii, V. M., Zyrianov, P. S., and Silin, V. P., Collision integral of charged particles in a magnetic field, *ZETF* **42**, 896 (1962) [*JETP* **15**, 619 (1962)].
151. Zhelezniakov, V. V., Solar and planetary radio emission, *UFN* **64**, 113 (1958) [CFSTI].
152. Zhelezniakov, V. V., On the question of nonlinear effects in a magnetoactive plasma, *Izv. Vuz, Radiofizika* **1**(5–6), 29 (1958) [CFSTI].
153. Zhelezniakov, V. V., On electromagnetic wave interaction in a plasma, *Izv. Vuz, Radiofizika* **1**(4), 32 (1958) [CFSTI].
154. Zhelezniakov, V. V., On synchrotron radiation and instability of a system of charged particles in a plasma, *Izv. Vuz, Radiofizika* **2**, 14 (1959) [CFSTI].
155. Zhelezniakov, V. V., On electromagnetic wave interaction in a plasma, *Izv. Vuz, Radiofizika* **2**, 858 (1959) [CFSTI].
156. Zhelezniakov, V. V., On instability of a magnetoactive plasma relative to high-frequency electromagnetic disturbances, *Izv. Vuz, Radiofizika* **4**, 619, 849 (1961) [CFSTI].
157. Zaslavskii, G. N. and Moiseev, S. S., On some peculiarities in the behavior

of a relativistic plasma with anisotropic electron velocity distribution, *ZETF* **42**, 1054 (1962) [*JETP* **15**, 731 (1962)].

158. Zakharov, V. E. and Karpman, V. I., On the nonlinear theory of plasma wave damping, *ZETF* **43**, 490 (1962) [*JETP* **16**, 351 (1963)].

159. Zubarev, D. N., On a new method in the theory of interacting particles, *ZETF* **25**, 548 (1953).

160. Zubarev, D. N., Two-time Green's functions in statistical physics, *UFN* **71**, 71 (1960) [*Uspekhi* **3**, 320 (1960)].

161. Zyrianov, P. S., Excitation spectrum of an electron plasma in a periodic ion field, *ZETF* **24**, 441 (1953) [ATS].

162. Zyrianov, P. S., On the influence of Coulomb correlations on the oscillation spectrum of an electron plasma, *ZETF* **34**, 232 (1958) [*JETP* **7**, 160 (1958)].

163. Zyrianov, P. S., Quantum theory of the excitation spectrum of an electron gas in a magnetic field, *ZETF* **40**, 1065 (1961) [*JETP* **13**, 751 (1961)].

164. Zyrianov, P. S., Quantum theory of acoustic oscillations of an electron–ion plasma in a magnetic field, *ZETF* **40**, 1353 (1961) [*JETP* **13**, 953 (1961)].

165. Zyrianov, P. S. and Kalashnikov, V. P. Quantum theory of the dielectric permittivity tensor of an electron plasma in a magnetic field, *ZETF* **41**, 1119 (1961) [*JETP* **14**, 799 (1962)].

166. Zyrianov, P. S. and Taluts, G. G., On electroacoustic phenomena in a degenerate electron–ion plasma, *ZETF* **36**, 145 (1959) [*JETP* **9**, 100 (1959)].

167. Imshennik, V. S., On shockwave structure in a high-temperature dense plasma, *ZETF* **42**, 236 (1962) [*JETP* **15**, 167 (1962)].

168. Kadomtsev, B. B., On a hydrodynamic description of plasma oscillations, *ZETF* **31**, 1083 (1956) [*JETP* **4**, 926 (1957)].

169. Kadomtsev, B. B., On fluctuations in a gas, *ZETF* **32**, 943 (1957) [*JETP* **5**, 771 (1957)].

170. Kadomtsev, B. B., On a field acting in a plasma, *ZETF* **33**, 151 (1957) [*JETP* **6**, 117 (1958)].

171. Kadomtsev, B. B., On low-pressure plasma hydrodynamics; "Plasma Physics and the Problem of Controlled Thermonuclear Reactions," Vol. 4. AN SSSR Press, Moscow, 1958 [PP].

172. Kadomtsev, B. B., On instability of a plasma in a magnetic field in the presence of ion beams; "Plasma Physics and the Problem of Controlled Thermonuclear Reactions," Vol. 4. AN SSSR Press, Moscow, 1958 [PP].

173. Kadomtsev, B. B., On the dynamics of a plasma in a strong magnetic field; "Plasma Physics and the Problem of Controlled Thermonuclear Reactions," Vol. 4. AN SSSR Press, Moscow, 1958 [PP].

174. Kadomtsev, B. B., On convective instability of plasma; "Plasma Physics and the Problem of Controlled Thermonuclear Reactions," Vol. 4; AN SSSR Press, Moscow, 1958 [PP].

175. Kadomtsev, B. B., On the convective instability of a plasma filament, *ZETF* **37**, 1096 (1959) [*JETP* **10**, 780 (1960)].

176. Kadomtsev, B. B., On the equilibrium of a plasma with spiral symmetry, *ZETF* **37**, 1352 (1959) [*JETP* **10**, 962 (1960)].

177. Kadomtsev, B. B., On stability of a low-pressure plasma, *ZETF* **37**, 1646 (1959) [*JETP* **10**, 1167 (1960)].

178. Kazantsev, A. P. and Gilinskii, A. A., On interaction of transverse oscillations in a plasma, *ZETF* **41**, 154 (1961) [*JETP* **14**, 112 (1961)].

179. Kazarinov, R. F. and Konstantinov, O. V., Dispersion theory of the high-frequency exciton conductivity of a crystal, *ZETF* **40**, 936 (1961) [*JETP* **13**, 654 (1961)].

180. Kaner, E. A., Cyclotron resonance in a plasma, *ZETF* **33**, 544 (1957) [*JETP* **6**, 425 (1958)].

181. Kaplan, S. S., On the Larmor theory of plasma, *ZETF* **36**, 1927 (1959) [*JETP* **9**, 1370 (1959)].

182. Kirochkin, Iu. A., On the theory of fluctuations in magnetohydrodynamics, *Izv. Vuz, Radiofizika* **5**, 1104 (1962) [CFSTI].

183. Kitsenko, A. B. and Stepanov, K. N., On instability of a plasma with an anisotropic ion and electron velocity distribution, *ZETF* **38**, 1840 (1960) [*JETP* **11**, 1323 (1960)].

184. Kitsenko, A. B. and Stepanov, K. N., On cyclotron instability in plasma, *ZTF* **31**, 176 (1961) [*Tech. Phys.* **6**, 127 (1961)].

185. Kitsenko, A. B. and Stepanov, K. N., On the passage of a beam of charged particles through a magnetoactive plasma, *Ukr. Fiz. Zh.* **6**, 297 (1961).

186. Klimontovich, Iu. L., Relativistic equation for the quantum distribution function, *DAN SSSR* **87**, 927 (1952).

187. Klimontovich, Iu. L., On correlation functions for quantum systems, *ZETF* **30**, 977 (1956) [*JETP* **3**, 781 (1956)].

188. Klimontovich, Iu. L., On the space-time correlation functions of a system of particles with electromagnetic interaction, *ZETF* **34**, 173 (1958) [*JETP* **7**, 119 (1958)].

189. Klimontovich, Iu. L., Energy losses by charged particles in excitation of oscillations in a plasma, *ZETF* **36**, 1405 (1959) [*JETP* **9**, 999 (1959)].

190. Klimontovich, Iu. L., Relativistic kinetic equations for a plasma, *ZETF* **37**, 735 (1959) [*JETP* **10**, 524 (1960)].

191. Klimontovich, Iu. L., Relativistic kinetic equation for a plasma, *ZETF* **38**, 1212 (1960) [*JETP* **11**, 876 (1960)].

192. Klimontovich, Iu. L. and Silin, V. P., On spectra of systems of interacting particles, *ZETF* **23**, 151 (1952) [CB].

193. Klimontovich, Iu. L. and Silin, V. P., On the spectra of systems of interacting particles and collective losses during passage of charged particles through a substance, *UFN* **70**, 247 (1960) [*Uspekhi* **3**, 84 (1960)].

194. Klimontovich, Iu, L. and Silin, V. P., On magnetohydrodynamics for a nonisothermal collisionless plasma, *ZETF* **40**, 1213 (1961) [*JETP* **13**, 852 (1961)].

195. Klimontovich, Iu. L. and Silin, V. P., On the theory of fluctuation of particle distributions in a plasma, *ZETF* **42**, 286 (1962) [*JETP* **15**, 199 (1962)].

196. Klimontovich, Iu. L. and Temko, S. V., Quantum kinetic equation for a plasma taking account of correlation, *ZETF* **33**, 132 (1957) [*JETP* **6**, 102 (1958)].

197. Klimontovich, Iu. L. and Ebeling, V., Hydrodynamic description of charged-particle motion in a weakly ionized plasma, *ZETF* **43**, 146 (1962) [*JETP* **16**, 104 (1963)].

198. Kovner, M. S., Kinetic analysis of charged particle flux interaction with a fixed plasma, *Izv. Vuz, Radiofizika* **3**, 631, 736 (1960) [CFSTI].
199. Kovner, M. S., On the question of the instability of low-frequency electromagnetic waves in a plasma penetrated by a charged-particle flux, *ZETF* **40**, 527 (1961) [*JETP* **13**, 369 (1961)].
200. Kovner, M. S., On a case of Cerenkov instability in a nonequilibrium magnetoactive plasma, *Izv. Vuz, Radiofizika* **4**, 765 (1961) [CFSTI].
201. Kovner, M. S., On a case of wave excitation in a nonequilibrium magnetoactive plasma, *Izv. Vuz, Radiofizika* **4**, 1035 (1961) [CFSTI].
202. Kovner, M. S., On the question of wave excitation in an unbounded nonequilibrium plasma, *ZTF* **32**, 145 (1962) [*Tech. Phys.* **7**, 101 (1962)].
203. Kovrizhnykh, L. M., On oscillations of a cylindrical cavity in a fully ionized plasma, *ZETF* **36**, 839 (1959) [*JETP* **9**, 592 (1959)].
204. Kovrizhnykh, L. M., Influence of inelastic collisions on the electron velocity distribution, *ZETF* **37**, 490 (1959) [*JETP* **10**, 347 (1960)].
205. Kovrizhnykh, L. M., Electron velocity distribution in a strong electric field, *ZETF* **37**, 1394 (1959) [*JETP* **10**, 989 (1960)].
206. Kovrizhnykh, L. M., On oscillations of an electron–ion plasma, *ZETF* **37**, 1692 (1959) [*JETP* **10**, 1198 (1960)].
207. Kovrizhnykh, L. M. and Rukhadze, A. A., On the instability of longitudinal oscillations of an electron–ion plasma, *ZETF* **38**, 850 (1960) [*JETP* **11**, 615 (1960)].
208. Kogan, V. I., On the rate of temperature equalization of charged particles in a plasma; "Plasma Physics and the Problem of Controlled Thermonuclear Reactions," Vol. 1. AN SSSR Press, Moscow, 1958 [PP].
209. Kogan, V. I., Spectral line broadening in a high-temperature plasma; "Plasma Physics and the Problem of Controlled Thermonuclear Reactions," Vol. 4. AN SSSR Press, Moscow, 1958 [PP].
210. Kogan, V. I., Fluctuating microfield and multiple collisions in a gas of charged particles, *DAN SSSR* **135**, 1374 (1960) [*Sov. Phys. Dokl.* **5**, 1316 (1961)].
211. Kogan, V. I. and Migdal, A. B., Dependence of the bremsstrahlung spectrum on the electron temperature of a plasma; "Plasma Physics and the Problem of Controlled Thermonuclear Reactions," Vol. 4. AN SSSR Press, Moscow, (1958) [PP].
212. Kolomenskii, A. A., On the electrodynamics of a gyrotropic medium, *ZETF* **24**, 167 (1953) [ATS].
213. Kolomenskii, A. A., Radiation of a uniformly moving electron in a plasma in a magnetic field, *DAN SSSR* **106**, 982 (1956) [*Sov. Phys. Dokl.* **1**, 133 (1956)].
214. Komarov, N. N. and Fadeev, V. M., Plasma in a self-consistent magnetic field, *ZETF* **41**, 528 (1961) [*JETP* **14**, 378 (1962)].
215. Konstantinov, O. V. and Perel', V. I., Quantum theory of the spatial dispersion of electric and magnetic susceptibilities, *ZETF* **37**, 786 (1959) [*JETP* **10**, 560 (1960)].
216. Konstantinov, O. V. and Perel', V. I., Graphical technique for evaluation of kinetic quantities, *ZETF* **39**, 197 (1960) [*JETP* **12**, 142 (1961)].
217. Konstantinov, O. V. and Perel', V. I., Particle collisions in a high-temperature plasma, *ZETF* **39**, 861 (1960) [*JETP* **12**, 597 (1961)].

218. Konstantinov, O. V. and Perel', V. I., Refinement of plasma kinetic coefficients, *ZETF* **41**, 1328 (1961) [*JETP* **14**, 944 (1962)].
219. Koniukov, M. V., Nonlinear Langmuir oscillations of electrons in a plasma, *ZETF* **37**, 799 (1959) [*JETP* **10**, 570 (1960)].
220. Kudriavtsev, V. S., Energetic diffusion of fast ions in an equilibrium plasma, *ZETF* **34**, 1558 (1958) [*JETP* **7**, 1075 (1958)].
221. Kudinov, E. K. and Pavlov, S. T., One-particle excitations in a nondegenerate electron gas, *ZETF* **42**, 839 (1962) [*JETP* **15**, 585 (1962)].
222. Kulik, I. O., On the momentum distribution function in a gas of Fermi particles in the high density limit, *ZETF* **40**, 1343 (1961) [*JETP* **13**, 946 (1961)].
223. Kulik, I. O., Peculiarities of the collective energy losses upon passage of fast electrons through an anisotropic plasma, *ZETF* **42**, 543 (1962) [*JETP* **15**, 380 (1962)].
224. Kurilko, V. I., On the kinetic theory of electromagnetic wave reflection from a moving plasma, *ZTF* **31**, 71 (1961) [*Tech. Phys.* **6**, 50 (1961)].
225. Kurilko, V. I., Electromagnetic wave reflection from a plasma moving in a slow-wave waveguide, *ZTF* **31**, 899 (1961) [*Tech. Phys.* **6**, 655 (1962)].
226. Kurilko, V. I. and Miroshnichenko, V. I., Plasma reflection of electromagnetic waves, *UFZh* **6**, 415 (1961).
227. Landau, L. D., Kinetic equation in the Coulomb interaction case, *ZETF* **7**, 203 (1937) [CFSTI].
228. Landau, L. D., On the oscillations of an electron plasma, *ZETF* **16**, 574 (1946) [CB].
229. Landau, L. D., Theory of a Fermi fluid, *ZETF* **30**, 1058 (1956) [*JETP* **3**, 920 (1957)].
230. Landau, L. D., Oscillations of a Fermi fluid, *ZETF* **32**, 59 (1957) [*JETP* **5**, 107 (1957)].
231. Landau, L. D., On the theory of a Fermi fluid, *ZETF* **35**, 97 (1958) [*JETP* **8**, 70 (1959)].
232. Landau, L. D. and Lifshitz, E. M., On hydrodynamic fluctuations, *ZETF* **32**, 618 (1957) [*JETP* **5**, 512 (1957)].
233. Larkin, A. I., Particle passage through a plasma, *ZETF* **37**, 264 (1959) [*JETP* **10**, 186 (1960)].
234. Larkin, A. I., Thermodynamic functions of a low-temperature plasma, *ZETF* **38**, 1896 (1960) [*JETP* **11**, 1363 (1960)].
235. Levin, M. L., On the electrodynamic theory of thermal radiation, *DAN SSSR* **102**, 53 (1955) [CFSTI].
236. Leontovich, M. A., Generalization of the Kramers-Kronig formulas to a medium with spatial dispersion, *ZETF* **40**, 907 (1961) [*JETP* **13**, 634 (1961)].
237. Leontovich, M. A. and Osovets, S. M., On the mechanism of current compression with a fast and powerful gas discharge, *AE* **3**, 81 (1956) [CB].
238. Leontovich, M. A. and Rytov, S. M., On the theory of electrical fluctuations, *DAN SSSR* **87**, 535 (1952).
239. Leontovich, M. A. and Rytov, S. M., On a differential law for the intensity of electrical fluctuations and the effect of skin-effect on it, *ZETF* **23**, 246 (1952).

240. Liperovskii, V. A. On anisotropy in the propagation of longitudinal electro-acoustic oscillations in a drifting plasma, *ZETF* **39**, 1363 (1960) [*JETP* **12**, 951 (1961)].

241. Lovetskii, E. E. and Rukhadze, A. A., On the hydrodynamics of a noniso-thermal plasma, *ZETF* **41**, 1845 (1961) [*JETP* **14**, 1312 (1962).

242. Luchina, A. A., On longitudinal oscillations of a plasma, *ZETF* **28**, 17 (1955) [*JETP* **1**, 12 (1955)].

243. Liubarskii, G. Ia., On the kinetic theory of shock waves, *ZETF* **40**, 1050 (1961) [*JETP* **13**, 740 (1961)].

244. Liubarskii, G. Ia., and Polovin, R. V., On splitting of unstable shock-waves in magnetohydrodynamics, *ZETF* **36**, 1272 (1959) [*JETP* **9**, 902 (1959)].

245. Milant'ev, V. P. and Popov, Iu. A., Thermal plasma fluctuations, *Vestnik MGU* Ser. 3, (4), 55 (1962).

246. Miller, M. A., Acceleration of plasma bunches by high-frequency electro-magnetic fields, *ZETF* **36**, 1909 (1959) [*JETP* **9**, 1358 (1959)].

247. Mikhailovskii, A. B., On the diamagnetic instability of a plasma with a large ion Larmor radius. *ZETF* **43**, 230 (1962) [*JETP* **16**, 164 (1963)].

248. Mikhailovskii, A. B., Convective instability and the effect of stabilization on a rarefied high-temperature plasma, *ZETF* **43**, 509 (1962) [*JETP* **16**, 364 (1963)].

249. Morozov, A. I., Cerenkov generation of magnetosonic waves; "Plasma Physics and the Problem of Controlled Thermonuclear Reactions," Vol. 4. AN SSSR Press, Moscow, 1958 [PP].

250. Morozov, A. I. and Solov'ev, L. S., On the extinction of plasma filament oscillations; "Plasma Physics and the Problem of Controlled Thermonuclear Reactions," Vol. 4, AN SSSR Press, Moscow, 1958 [PP].

251. Morozov, A. I., and Solov'ev, L. S., Kinetic analysis of some equilibrium plasma configurations, *ZETF* **40**, 1316 (1961) (*JETP* **13**, 927 (1961)].

252. Nezlin, M. V., Electrostatic instability of an intense electron beam in a plasma, *ZETF* **41**, 1015 (1961) [*JETP* **14**, 723 (1962)].

253. Nekrasov, F. M., On the nonlinear theory of stationary processes in an electron plasma, *ZETF* **38**, 233 (1960) [*JETP* **11**, 170 (1960)].

254. Nekrasov, F. M., On the initial stage of development of perturbations in a plasma in an external magnetic field, *ZETF* **43**, 483 (1962) [*JETP* **16**, 346 (1962)].

255. Osovets, S. M., On holding a plasma by means of a traveling magnetic field; "Plasma Physics and the Problem of Controlled Thermonuclear Reactions," Vol. 4. AN SSSR Press, Moscow, 1958 [PP].

256. Osovets, S. M., Dynamic stabilization of a plasma turn, *ZETF* **39**, 311 (1960) [*JETP* **12**, 221 (1961)].

257. Pargamanik, L. E., On the kinetic theory of an electron gas in the presence of boundaries, *ZETF* **33**, 251 (1957) [*JETP* **6**, 154 (1958)].

258. Pargamanik, L. E., On the shift of atom energy levels in a plasma, *ZETF* **41**, 1112 (1961), [*JETP* **14**, 794 (1962)].

259. Perel', V. I. and Eliashberg, G. M., Electromagnetic wave absorption in plasma, *ZETF* **41**, 886 (1961) [*JETP* **14**, 633 (1962)].

260. Pistunovich, V. I. and Shafranov, V. D., Cyclotron radiation of ions in a plasma, *Nuclear Fusion* **1**, 189 (1961) [CFSTI].

261. Pitaevskii, L. P. and Kresin, V. Z., On the question of perturbations occurring during body motion in a plasma, *ZETF* **40**, 271 (1961) [*JETP* **13**, 185 (1961)].

262. Polovin, R. V., On the nonlinear theory of longitudinal plasma oscillations, *ZETF* **31**, 354 (1956) [*JETP* **4**, 290 (1957)].

263. Polovin, R. V., On the theory of simple magnetohydrodynamic waves, *ZETF* **39**, 463 (1960), [*JETP* **12**, 326 (1961)].

264. Polovin, R. V., Shockwaves in magnetohydrodynamics, *UFN* **72**, 33 (1960) [*Uspekhi* **3**, 677 (1961)].

265. Polovin, R. V. and Liubarskii, G. Ia., Impossibility of rarefaction shockwaves in magnetohydrodynamics, *ZETF* **35**, 510 (1958) [*JETP* **8**, 351 (1959)], *UFZh* **3**, 571 (1958) [CFSTI].

266. Polovin, R. V. and Tsintsadze, N. L., On small oscillations of an electron beam, *ZTF* **27**, 1466 (1957) [*Tech. Phys.* **2**, 1354 (1958)].

267. Polovin, R. V. and Tsintsadze, N. L., Circular waves in an electron–ion beam, *ZETF* **34**, 637 (1958) [*JETP* **7**, 440 (1958)].

268. Polovin, R. V. and Tsintsadze, N. L., On the theory of longitudinal oscillations of an electron–ion beam, *ZTF* **29**, 831 (1959) [*Tech. Phys.* **4**, 751 (1960)].

269. Polovin, R. V. and Cherkasova, K. P., On the magnetohydrodynamic description of a plasma, *ZTF* **32**, (1962) [*Tech. Phys.* **7**, 475 (1963)].

270. Prokhoda, I. G., On the theory of electromagnetic wave scattering by electron–density fluctuations in a plasma, *Uchen. Zapisk. Khar'k. Gos. Univ.* **98**, 61 (1958).

271. Piatigorskii, L. M., Electromagnetic waves in a plasma in a magnetic field, *Uchen. Zapisk. Khar'k. Gos. Univ.* **64**, 23 (1955).

272. Rapoport, V. O., On the growth of electromagnetic waves in a flow moving in a plasma in the presence of a magnetic field, *Izv. Vuz, Radiofizika* **3**, 737 (1960) [CFSTI].

273. Romanov, Iu. A. and Filippov, G. F., Interaction of fast electron fluxes with longitudinal plasma waves, *ZETF* **40**, 123 (1961) [*JETP* **13**, 87 (1961)].

274. Rudakov, L. I. and Sagdeev, R. Z., On a quasihydrodynamic description of a rarefied plasma in a magnetic field; "Plasma Physics and the Problem of Controlled Thermonuclear Reactions," Vol. 3. AN SSSR Press, Moscow, 1958 [PP].

275. Rudakov, L. I. and Sagdeev, R. Z., Investigation of the stability of a cylindrical plasma filament by the kinetic equation method; "Plasma Physics and the Problem of Controlled Thermonuclear Reactions," Vol. 4. AN SSSR Press, Moscow, 1958 [PP].

276. Rudakov, L. I. and Sagdeev, R. Z., Oscillations of an inhomogeneous plasma in a magnetic field, *ZETF* **37**, 1337 (1959) [*JETP* **10**, 952 (1960)].

277. Rukhadze, A. A. and Silin, V. P., On the magnetic susceptibility of a relativistic electron gas, *ZETF* **38**, 645 (1960) [*JETP* **11**, 463 (1960)].

278. Rukhadze, A. A. and Silin, V. P., Synchrotron absorption line shape in a plasma, *ZTF* **32**, 423 (1962) [*Tech. Phys.* **7**, 307 (1962)].

279. Rukhadze, A. A. and Silin, V. P., Linear electromagnetic phenomena in plasma, *UFN* **76**, 79 (1962), [*Uspekhi* **5**, 37 (1962)].

280. Rytov, S. M., "Theory of Electrical Fluctuations and Thermal Radiation," AN SSSR Press, Moscow, 1953 [CFSTI].

281. Rytov, S. M., Correlation theory of thermal fluctuations in an isotropic medium, *ZETF* **33**, 166 (1957) [*JETP* **6**, 130 (1958)].

282. Rytov, S. M., Correlation theory of Rayleigh light scattering, *ZETF* **33**, 514, 669 (1957) [*JETP* **6**, 401, 513(1958)].

283. Sagdeev, R. Z., On nonlinear motions of a rarefied plasma in a magnetic field; "Plasma Physics and the Problem of Controlled Thermonuclear Reactions," Vol. 4. AN SSSR Press, Moscow, 1958 [PP].

284. Sagdeev, R. Z., On absorption of electromagnetic waves propagated along a magnetic field in a plasma; "Plasma Physics and the Problem of Controlled Thermonuclear Reactions," Vol. 4. AN SSSR Press, Moscow, 1958 [PP].

285. Sagdeev, R. Z. and Shafranov, V. D., Oscillations of a plasma filament taking account of thermal ion motion; "Plasma Physics and the Problem of Controlled Thermonuclear Reactions," Vol. 4. AN SSSR Press, Moscow, 1958 [PP].

286. Sagdeev, R. Z. and Shafranov, V. D., Energy absorption of a high-frequency magnetic field in a high-temperature plasma; "Proc. Second Intern. Conf. Peaceful Uses Atomic Energy," Moscow, Atomizdat, 1959 [CB].

287. Sagdeev, R. Z. and Shafranov, V. D., On instability of a plasma with an anisotropic velocity distribution in a magnetic field, *ZETF* **39**, 181 (1960) [*JETP* **12**, 130 (1961)].

288. Saltanov, N. V. and Tkalich, V. S., Magnetohydrodynamic waves of finite amplitude, *ZTF* **30**, 1235 (1960) [*Tech. Phys.* **5**, 1188 (1961)].

289. Silin, V. P., On the excitation spectrum of a system of electrons and ions, *ZETF* **23**, 649 (1952).

290. Silin, V. P., Investigation of the spectrum of a system of many particles by the quantum kinetic equation method, *Transactions, Lebedev Physics Inst.* **6**, 199 (1955) [CFSTI].

291. Silin, V. P., On the theory of a degenerate electron fluid, *ZETF* **33**, 495 (1957) [*JETP* **6**, 387 (1958)].

292. Silin, V. P., On the theory of plasma waves of a degenerate electron fluid, *ZETF* **34**, 781 (1958) [*JETP* **7**, 538 (1958)].

293. Silin, V. P., Oscillations of a degenerate electron fluid, *ZETF* **35**, 1243 (1958) [*JETP* **8**, 870 (1959)].

294. Silin, V. P., On collective losses of fast electrons during passage through a substance, *ZETF* **37**, 273 (1959) [*JETP* **10**, 192 (1960)].

295. Silin, V. P., Electromagnetic fluctuations in media with spatial dispersion, *Izv. Vuz, Radiofizika* **2**, 198 (1959) [CFSTI].

296. Silin, V. P., On the electromagnetic properties of a relativistic plasma, *ZETF* **38**, 1576 (1960) [*JETP* **11**, 1136 (1960)].

297. Silin, V. P., Kinetic equation for rapidly varying processes, *ZETF* **38**, 1771 (1960), [*JETP* **11**, 1271 (1960)].

298. Silin, V. P., On the initial problem for a longitudinal field in a degenerate electron gas, *FMM* **10**, 942 (1960) [*Physics of Metals and Metallography* **10**(6), 147 (1960)].

299. Silin, V. P., On the electromagnetic properties of a relativistic plasma, *ZETF* **40**, 616 (1961) [*JETP* **13**, 430 (1961)].

300. Silin, V. P., On the collision integral for charged particles, *ZETF* **40**, 1768 (1961) [*JETP* **13**, 1244 (1961)].

301. Silin, V. P., On the high-frequency dielectric permittivity, *ZETF* **41**, 861 (1961) [*JETP* **14**, 617 (1962)].

302. Silin, V. P., On the theory of electromagnetic fluctuations in a plasma, *ZETF* **41**, 969 (1961) [*JETP* **14**, 689 (1962)].

303. Silin, V. P., Relativistic kinetic equation for rapidly varying processes in an ionized gas, *Izv. Vuz, Radiofizika* **4**, 1029 (1961) [CFSTI].

304. Silin, V. P. and Rukhadze, A. A., "Electromagnetic Properties of Plasma and Plasmalike Media," Atomizdat, Moscow, 1961 [GB (1965)].

305. Silin, V. P. and Fetisov, E. P., On electromagnetic properties of a relativistic plasma, *ZETF* **41**, 159 (1961) [*JETP* **14**, 115 (1962)].

306. Sitenko, A. G., On the passage of a charged particle through a magnet, *DAN SSSR* **98**, 377 (1954).

307. Sitenko, A. G., On passage of a charged particle through a lossy dielectric, *Uchen. Zapisk. Khar'k. Univ.* **64**, 17 (1955).

308. Sitenko, A. G. and Kaganov, M. I., On energy losses of a charged particle moving in an anisotropic medium, *DAN SSSR* **100**, 681 (1955).

309. Sitenko, A. G. and Kirochkin, Iu. A., On wave excitation in a plasma, *ZTF* **29**, 801 (1959) [*Tech. Phys.* **4**, 723 (1960)].

310. Sitenko, A. G. and Kirochkin, Iu. A., On hydromagnetic wave scattering by turbulent fluctuations; "Questions of Magnetohydrodynamics and Plasma Dynamics," Riga, 1959 [CFSTI].

311. Sitenko, A. G. and Kirochkin, Iu. A., On electromagnetic wave scattering by fluctuations in a plasma in the presence of a magnetic field, *Izv. Vuz, Radiofizika* **6** (1963) [CFSTI].

312. Sitenko, A. G. and Kirochkin, Iu. A., On hydromagnetic wave scattering and transformation by fluctuations, *ZTF* **33** (1963) [*Tech. Phys.* **8**, 1008 (1964)].

313. Sitenko, A. G. and Kolomenskii, A. A., On charged particle motion in an optically active anisotropic medium, *ZETF* **30**, 511 (1956) [*JETP* **3**, 410 (1956)].

314. Sitenko, A. G. and Simenog, I. V., On fluctuations in a degenerate electron gas, *Izv. Vuz, Radiofizika* **6**, 54 (1963) [CFSTI].

315. Sitenko, A. G. and Simenog, I. V., On the theory of fluctuations in superconductors, *UFZh* **8** (1963).

316. Sitenko, A. G. and Stepanov, K. N., On oscillations of an electron plasma in a magnetic field, *ZETF* **31**, 642 (1956) [*JETP* **4**, 512 (1957)].

317. Sitenko, A. G. and Stepanov, K. N., On charged particle interaction with an electron plasma, *Uchen. Zapisk. Khar'k. Univ.* **8**, 5 (1958).

318. Sitenko, A. G. and Tsien, Yu-tai, On dynamic friction and diffusion coefficients in a plasma, *ZTF* **32**, 1325 (1962) [*Tech. Phys.* **7**, 978 (1963)].

319. Skobov, V. G., On the theory of the conductivity of an electron gas in a strong magnetic field, *ZETF* **38**, 1304 (1960) [*JETP* **11**, 941 (1960)].

320. Soluian, S. I., Magnetosonic waves in a cylindrical plasma filament taking

account of nonlinearity and absorption, *ZETF* **43**, 185 (1962) [*JETP* **16**, 132 (1963)].

321. Stepanov, K. N., Kinetic theory of magnetohydrodynamic waves, *ZETF* **34**, 1292 (1958), [*JETP* **7**, 892 (1958)].

322. Stepanov, K. N., On electromagnetic wave attenuation in a plasma in a magnetic field, *ZETF* **35**, 283 (1958) [*JETP* **8**, 195 (1959)].

323. Stepanov, K. N., Low-frequency plasma oscillations in a magnetic field, *ZETF* **35**, 1155 (1958) [*JETP* **8**, 808 (1958)].

324. Stepanov, K. N., On electromagnetic wave penetration into a plasma, *ZETF* **36**, 1457 (1959), [*JETP* **9**, 1035 (1959)].

325. Stepanov, K. N., Magnetosonic waves in a rarefied plasma, *UFZh* **4**, 678 (1959).

326. Stepanov, K. N., On cyclotron absorption of electromagnetic waves in a plasma, *ZETF* **38**, 265 (1960) [*JETP* **11**, 192 (1960)].

327. Stepanov, K. N., On cyclotron resonance in a plasma; "Plasma Physics and Problems of Controlled Thermonuclear Fusion," AN UkrSSR Press, Kiev, 1962.

328. Stepanov, K. N., On the influence of thermal electron and ion motion on quasi-transverse radiowave propagation in a plasma; "Plasma Physics and Problems of Controlled Thermonuclear Fusion," AN UkrSSR Press, Kiev, 1962.

329. Stepanov, K. N. and Kitsenko, A. B., On electromagnetic wave excitation in a magnetoactive plasma by a charged particle beam, *ZTF* **31**, 167 (1961) [*Tech. Phys.* **6**, 120 (1961)].

330. Stepanov, K. N. and Pakhomov, V. I., On synchrotron radiation of a bounded plasma, *ZETF* **38**, 1564 (1960) [*JETP* **11**, 1126 (1960)].

331. Stepanov, K. N. and Tkalich, V. S., On electron plasma oscillations in a magnetic field, *ZTF* **28**, 1789 (1958) [*Tech. Phys.* **3**, 1649 (1959)].

332. Syrovatskii, S. I., Magnetohydrodynamics, *UFN* **62**, 247 (1957) [CFSTI].

333. Tamm, I. E., Theory of the magnetic thermonuclear reactor; "Plasma Physics and the Problem of Controlled Thermonuclear Reactions," Vol. 1. AN SSSR Press, Moscow, 1958 [PP].

334. Tamm, I. E. and Frank, I. M., Coherent radiation of a fast electron in a medium, *DAN SSSR* **14**, 107 (1937).

335. Tverskoi, B. A., On one-dimensional self-similar waves propagated in a plasma along a magnetic field, *ZETF* **42**, 833 (1962) [*JETP* **15**, 581 (1962)].

336. Temko, S. V., On the derivation of the Fokker-Planck equation for a plasma, *ZETF* **31**, 1021 (1956) [*JETP* **4**, 898 (1957)].

337. Temko, S. V., On the quantum kinetic equation for multisort systems of charged particles, *ZETF* **34**, 523 (1958) [*JETP* **7**, 361 (1958)].

338. Timofeev, A. V., Swinging of ion acoustic oscillations in an anisotropic plasma, *ZETF* **39**, 397 (1960) [*JETP* **12**, 281 (1961)].

339. Tkalich, V. S., Finite amplitude waves in a multicomponent conducting medium, *ZETF* **39**, 73 (1960) [*JETP* **12**, 52 (1961)].

340. Tkalich, V. S. and Saltanov, N. V., On finite amplitude waves in nonideal magnetohydrodynamics, *ZTF* **31**, 1231 (1961) [*Tech. Phys.* **6**, 896 (1962)].

341. Tkalich, V. S. and Saltanov, N. V., On nonlinear Langmuir oscillations, *ZTF* **32**, 156 (1962) [*Tech. Phys.* **7**, 109 (1962)].

342. Tolmachev, V. V., On finding time correlation functions for statistical systems with long-range effects in times to the mean-free-path time, *DAN SSSR* 113, 301 (1957) [*Sov. Phys. Dokl.* 2, 124 (1957)].

343. Tolmachev, V. V., Time correlations in classical statistical systems consisting of a large number of interacting particles, *DAN SSSR* 112, 842 (1957) [*Sov. Phys. Dokl.* 2, 85 (1957)].

344. Tolmachev, V. V. and Tiablikov, S. V., On the classical theory of strong electrolytes, *DAN SSSR* 119, 314 (1958) [CB] [*Dokl. Phys. Chem.*].

345. Trubnikov, B. A., Plasma radiation in a magnetic field, *DAN SSSR* 118, 913 (1958) [*Sov. Phys. Dokl.* 3, 136 (1958)].

346. Trubnikov, B. A., Reduction of the kinetic equation to differential form in the case of Coulomb collisions, *ZETF* 34, 1341 (1958) [*JETP* 7, 926 (1958)].

347. Trubnikov, B. A., Electromagnetic waves in a relativistic plasma in the presence of a magnetic field; "Plasma Physics and the Problem of Controlled Thermonuclear Reactions," Vol. 3. AN SSSR Press, Moscow, 1958 [PP].

348. Trubnikov, B. A., Relation between the radiation absorption and emission coefficients for a plasma in a magnetic field; "Plasma Physics and the Problem of Controlled Thermonuclear Reactions," Vol. 4. AN SSSR Press, Moscow, 1958 [PP].

349. Trubnikov, B. A., Plasma behavior in a rapidly varying magnetic field; "Plasma Physics and the Problem of Controlled Thermonuclear Reactions," Vol. 4. AN SSSR Press, Moscow, 1958 [PP].

350. Trubnikov, B. A. and Bazhanova, A. E., Magnetic radiation of a plasma layer; "Plasma Physics and the Problem of Controlled Thermonuclear Reactions," Vol. 3. AN SSSR Press, Moscow, 1958 [PP].

351. Trubnikov, B. A. and Kudriavtsev, V. S., Plasma radiation in a magnetic field; "Proc. Second Conf. Peaceful Uses of Atomic Energy, in Geneva," Moscow, Atomizdat, 1959 [CB].

352. Tiablikov, S. V. and Tolmachev, V. V., Distribution functions for a classical electron gas, *DAN SSSR* 114, 1210 (1957) [*Sov. Phys. Dokl.* 2, 299 (1957)].

353. Fainberg, Ia. B., Particle acceleration in a plasma, *AE* 6, 431 (1959) [CB] [*Sov. J. Atomic Energy* 6, 297 (1960)].

354. Fainberg, Ia. B., Interaction of charged particle beams with a plasma, *AE* 11, 313 (1961) [CB].

355. Fainberg, Ia. B. and Gorbatenko, M. F., Electromagnetic waves in a plasma in a magnetic field, *ZTF* 29, 549 (1959) [*Tech. Phys.* 4, 487 (1959)].

356. Fainberg, Ia. B., Gorbatenko, M. F., and Kurilko, V. I., Vavilov-Cerenkov radiation in a bounded isotropic medium; "Plasma Physics and Problems of Controlled Thermonuclear Fusion," AN UkrSSR Press, Kiev, 1962.

357. Fainberg, Ia. B. and Kurilko, V. I., Electromagnetic pressure on a charge moving in a magnetic field, *ZTF* 29, 939 (1959) [*Tech. Phys.* 4, 855 (1960)].

358. Fainberg, Ia. B., Kurilko, V. I., and Shapiro, V. D., On the question of the nature of the instability during interaction between charged-particle beams and a plasma, *ZTF* 31, 639 (1961) [*Tech. Phys.* 6, 454 (1962)].

359. Fainberg, Ia. B., Nekrasov, F. M., and Kurilko, V. I., On the theory of nonlinear longitudinal waves in a plasma; "Plasma Physics and Problems of Controlled Thermonuclear Fusion," AN UkrSSR Press, Kiev, 1962.

226   Bibliography

360. Fainberg, Ia. B. and Khizhniak, N. A., Charge density waves in modulated beams; "Plasma Physics and Problems of Controlled Thermonuclear Fusion," AN UkrSSR Press, Kiev, 1962.

361. Fainberg, Ia. B. and Shapiro, V. D., Waveguide properties of a plasma cylinder in a longitudinal magnetic field taking account of thermal motion in the plasma; "Plasma Physics and Problems of Controlled Thermonuclear Fusion," AN UkrSSR Press, Kiev, 1962.

362. Feinberg, E. L., Collective oscillations of electrons in a crystal, *ZETF* **34**, 1125 (1958) [*JETP* **7**, 780 (1958)].

363. Fok, V. A., "Radiowave Diffraction around the Globe." AN SSSR Press, Moscow, 1946 [MDF].

364. Fradkin, E. S., On the theory of transport processes in a plasma in a magnetic field, *ZETF* **32**, 1176 (1957) [*JETP* **5**, 596 (1957)].

365. Frank, I. M., On the critical velocity during light radiation in optically anisotropic media, *ZETF* **38**, 1751 (1960) [*JETP* **11**, 1263 (1960)].

366. Frank-Kamenetskii, D. A., On the proper oscillations of a bounded plasma, *ZETF* **39**, 669 (1960) [*JETP* **12**, 469 (1961)].

367. Frank-Kamenetskii, D. A., "Plasma—Fourth State of Matter." Atomizdat, Moscow, 1961 [CB].

368. Khizhniak, N. A., On the theory of nonlinear processes in a one-component plasma; "Plasma Physics and Problems of Controlled Thermonuclear Fusion," AN UkrSSR Press, Kiev, 1962.

369. Tserkovnikov, Iu. A., Plasma stability in a strong magnetic field, *ZETF* **32**, 67 (1957) [*JETP* **5**, 58 (1957)].

370. Tsintsadze, N. L., Determination of the relativistic electron beam shape, *ZTF* **29**, 24 (1959) [*Tech. Phys.* **4**, 21 (1959)].

371. Tsytovich, V. N., On spatial dispersion in a relativistic plasma, *ZETF* **40**, 1775 (1961) [*JETP* **13**, 1249 (1960)].

372. Tsytovich, V. N., Macroscopic renormalization of the mass and energy losses of charged particles in a medium, *ZETF* **42**, 457 (1962) [*JETP* **15**, 320 (1962)].

373. Tsytovich, V. N., On energy losses of charged particles in a plasma, *ZETF* **42**, 803 (1962) [*JETP* **15**, 561 (1962)].

374. Tsytovich, V. N., On the effect of radiation on a charged particle passing through a magnetoactive plasma, *ZETF* **43**, 327 (1962) [*JETP* **16**, 234 (1963)].

375. Shabanskii, V. P., Structure of the transition layer between a plasma and a magnetic field, *ZETF* **40**, 1059 (1961) [*JETP* **13**, 746 (1961)].

376. Shapiro, V. D., On the influence of electrostatic instability on plasma electrical conductivity and temperature, *Izv. Vuz, Radiofizika* **4**, 867 (1961) [CFSTI].

377. Shapiro, V. D., On the nonlinear theory of a plasma in a magnetic field; "Plasma Physics and Problems of Controlled Thermonuclear Fusion," AN UkrSSR Press, Kiev, 1962.

378. Shapiro, V. D., Plasma heating by the acoustic resonance method; "Plasma Physics and Problems of Controlled Thermonuclear Fusion," AN UkrSSR Press, Kiev, 1962.

379. Shapiro, V. D. and Shevchenko, V. I., On the nonlinear theory of charged-

particle beam interaction with a plasma in a magnetic field, *ZETF* **42**, 1515 (1962) [*JETP* **15**, 1053 (1962)].

380. Shafranov, V. D., On the stability of a cylindrical gas conductor in a magnetic field, *AE* **5**, 38 (1956) [CB].

381. Shafranov, V. D., Shockwave structure in a plasma, *ZETF* **32**, 1454 (1957) [*JETP* **5**, 1183 (1957)].

382. Shafranov, V. D., On equilibrium magnetohydrodynamic configurations, *ZETF* **33**, 710 (1957) [*JETP* **6**, 545 (1958)].

383. Shafranov, V. D., Magnetic vortex rings, *ZETF* **33**, 831 (1957) [*JETP* **6**, 642 (1958)].

384. Shafranov, V. D., Electromagnetic wave propagation in a medium with spatial dispersion, *ZETF* **34**, 1475 (1958) [*JETP* **7**, 1019 (1958)].

385. Shafranov, V. D., On the stability of a plasma filament in the presence of a longitudinal magnetic field and a conducting housing; "Plasma Physics and the Problem of Controlled Thermonuclear Reactions," Vol. 2. AN SSSR Press, Moscow, 1958 [PP].

386. Shafranov, V. D., On the stability of a plasma filament with distributed current; "Plasma Physics and the Problem of Controlled Thermonuclear Reactions," Vol. 4. AN SSSR Press, Moscow, 1958 [PP].

387. Shafranov, V. D., On the derivation of the plasma dielectric permittivity tensor; "Plasma Physics and the Problem of Controlled Thermonuclear Reactions," Vol. 4. AN SSSR Press, Moscow, 1958 [PP].

388. Shafranov, V. D., Plasma refractive index in a magnetic field in the cyclotron resonance domain; "Plasma Physics and the Problem of Controlled Thermonuclear Reactions," Vol. 4. AN SSSR Press, Moscow, 1958 [PP].

389. Shafranov, V. D., On the equilibrium of a plasma torus in a magnetic fiield, *ZETF* **37**, 1088 (1959) [*JETP* **10**, 775 (1960)].

390. Shafranov, V. D., "Electromagnetic Waves in Plasmas," Moscow, IAE AN SSSR, 1960.

391. Eidman, V. Ia., Radiation of an electron moving in a magnetoactive plasma, *ZETF* **34**, 131 (1958); **36**, 1335 (1959) [*JETP* **7**, 91 (1958); **9**, 947 (1959)].

392. Eidman, V. Ia., On the radiation reactive force in a magnetoactive plasma, *Izv. Vuz, Radiofizika* **3**, 192 (1960) [CFSTI].

393. Eidman, V. Ia., On transient radiation on a plasma boundary taking account of spatial dispersion, *Izv. Vuz, Radiofizika* **5**, 478 (1962), [CFSTI].

394. Eidman, V. Ia., Radiation of a plasma wave by a charge moving in a magnetoactive plasma, *ZETF* **41**, 1971 (1961) [*JETP* **14**, 1401 (1962)].

395. Eliashberg, G. M., Kinetic equation for a degenerate system of Fermi particles, *ZETF* **41**, 1241 (1961) [*JETP* **14**, 886 (1962)].

396. Iakovenko, V. M., Transient radiation in a plasma taking account of temperature, *ZETF* **41**, 385 (1961) [*JETP* **14**, 278 (1962)].

397. Adams, E. and Holstein, T., Quantum theory of transverse galvanomagnetic phenomena, *J. Phys. Chem. Solids* **10**, 254 (1959).

398. Akhiezer, A. I., On electromagnetic wave interaction with charged particles and the oscillations of an electron plasma, *Nuovo Cimento* **4** (3), 591 (1956).

399. Alfven, H., On the existence of electromagnetic-hydrodynamic waves, *Arkiv mat. ast. fys.* **29B** (2) (1942); *Nature* **150**, 405 (1942).

400. Alfven, H., "Cosmical Electrodynamics." Oxford Univ. Press, London, 1950.
401. Amer, S., Nonlinear theory of oscillations and waves in plasma, *J. Electr. Contr.* **5**, 105 (1958).
402. Anderson, N., Plasma oscillations in a permanent magnetic field, *Proc. Phys. Soc.* **77**, 971 (1961).
403. Astrom, E., Waves in an ionized plasma, *Arkiv Fys.* **2**, 443 (1950).
404. Auer, P. L., Instability of longitudinal oscillations in plasma, *Phys. Rev. Letters* **1**, 411 (1958).
405. Auer, P. L., Hurwitz, H., and Miller, R. D., Collective oscillations in a cold plasma, *Phys. Fluids* **1**, 501 (1958).
406. Auer, P. L., Hurwitz, H., and Kilb, R. W., Magnetic compression waves at low Mach number in a collisionless plasma, *Phys. Fluids* **4**, 1105 (1961).
407. Balescu, R., Irreversible processes in an ionized plasma, *Phys. Fluids* **3**, 52 (1960).
408. Balescu, R., Equilibrium build-up in plasma, *J. Nucl. Energy* **C2**, 169 (1961).
409. Balescu, R. and Taylor, H. S., Binary correlations in ionized gases, *Phys. Fluids* **4**, 85 (1961).
410. Balescu, R., Equilibrium build-up in a quantum plasma, *Phys. Fluids* **4**, 94 (1961).
411. Banos, A., Fundamental wave functions in an unbounded magnetohydrodynamic field, *Phys. Rev.* **97**, 1435 (1955).
412. Banos, A., Magnetohydrodynamic waves in incompressible and compressible fluids, *Proc. Roy. Soc. (London)* **A233**, 350 (1955).
413. Bardasis, A. and Schrieffer, J., Excitons and plasmons in superconductors, *Phys. Rev.* **121**, 1050 (1961).
414. Beard, D. B., Microwave emission by a high-temperature plasma, *Phys. Rev. Letters* **2**, 81 (1959).
415. Beard, D. B., Cyclotron radiation in plasma, *Phys. Fluids* **2**, 379 (1959).
416. Beard, D. B., Relativistic analysis of cyclotron radiation in a hot plasma, *Phys. Fluids* **3**, 324 (1960).
417. Beard, D. B., Optical properties and radiation of a relativistic plasma near cyclotron resonance, *J. Nucl. Energy* **C2**, 94 (1961).
418. Bekefi, G., Hirshfield, J. L., and Brown, S. C., Cyclotron radiation in a plasma with non-Maxwellian distribution, *Phys. Rev.* **122**, 1037 (1961).
419. Bekefi, G., Hirshfield, J. L., and Brown, S. C., Kirchhoff radiation law in a plasma with non-Maxwellian distribution, *Phys. Fluids* **4**, 173 (1961).
420. Bennett, W. H., Self-focussing beams, *Phys. Rev.* **98**, 1584 (1955).
421. Berger, J. M., Newcomb, W. A., Dawson, J. M., Frieman, E. A., Kulsrud, R. M., and Lenard, A., Heating and containment of a plasma by a high-frequency electromagnetic field, *Phys. Fluids* **1**, 301 (1958).
422. Bernstein, I., Greene, J., and Kruskal, M., Nonlinear plasmon oscillations, *Phys. Rev.* **108**, 546 (1957).
423. Bernstein, I. B., Waves in a plasma in a magnetic field, *Phys. Rev.* **109**, 10 (1958).
424. Bernstein, I. B., Plasma oscillations perpendicular to a permanent magnetic field, *Phys. Fluids* **3**, 489 (1960).

425. Bernstein, I. B., Frieman, E. A., Kruskal, M. D., and Kulsrud, R. M., Energy principle hydromagnetic stability problems, *Proc. Roy. Soc. (London)* **A244**, 217 (1958).

426. Bernstein, I. B., Frieman, E. A., Kulsrud, R. M., and Rosenbluth, M. N., Ionic waves instability, *Phys. Fluids* **3**, 136 (1960).

427. Bernstein, I. B., Greene, J. M., and Kruskal, M. D., Exact theory of nonlinear plasma oscillations, *Phys. Rev.* **108**, 546 (1957).

428. Bernstein, I. B. and Kulsrud, R. M., Ion wave instability, *Phys. Fluids* **3**, 937 (1960).

429. Bernstein, I. B. and Trehan, S., Oscillations in a plasma, *Nuclear Fusion* **1**, 3 (1960).

430. Berz, F., On the theory of waves in plasma, *Proc. Phys. Soc.* **69**, 939 (1956).

431. Bohm, D. and Gross, E. P., Theory of oscillations in plasmas. 1. Nature of plasma waves, *Phys. Rev.* **75**, 1851 (1949).

432. Bohm, D. and Gross, E. P., Theory of oscillations in plasmas. 2. Excitation and damping of oscillations, *Phys. Rev.* **75**, 1864 (1949).

433. Bohm, D. and Gross, E. P., Influence of boundaries on plasma oscillations, *Phys. Rev.* **79**, 992 (1950).

434. Bohm, D. and Pines, D., Collective description of electron interaction. 1. Magnetic interaction, *Phys. Rev.* **82**, 625 (1951).

435. Bohm, D. and Pines, D., Collective description of electron interaction. 2. Coulomb interaction in a degenerate electron gas, *Phys. Rev.* **92**, 609 (1953).

436. Bohr, N., Passage of atomic particles through a substance, *Dan. Mat. Fys. Medd.* **18**(8) (1948).

437. Bowles, K., Observation of radiowave scattering from the ionosphere, *Phys. Rev. Letters* **1**, 454 (1958).

438. Brittin, W. E., Statistical theory of transport phenomena in a fully ionized gas, *Phys. Rev.* **106**, 943 (1957).

439. Brout, R., Phonons and plasmons, *J. Nuclear Energy* **C2**, 46 (1961).

440. Brueckner, K., Fukuda, N., Sawada, K., and Brout, R., Correlation energy of a high-density electron gas: plasma oscillations, *Phys. Rev.* **108**, 507 (1957).

441. Brueckner, K., and Gell-Mann, M., Correlation energy of a high-density electron gas, *Phys. Rev.* **106**, 364 (1957).

442. Brueckner, K. and Watson, K., Use of the Boltzmann equation to study low density ionized gases. 2, *Phys. Rev.* **102**, 19 (1956).

443. Buchsbaum, S. J., Mower, L., and Brown, S. C., Interaction between plasma and electromagnetic waves, *Phys. Fluids* **3**, 806 (1960).

444. Budini, P. and Taffara, L., On energy losses by a relativistic particle in a polarized medium, *Nuovo Cimento* **4**, 23 (1956).

445. Buneman, O., Transverse waves in a plasma and plasma eddies, *Phys. Rev.* **112**, 1504 (1959).

446. Buneman, O., Current dissipation in an ionized medium, *Phys. Rev.* **115**, 503 (1959).

447. Buneman, O., Equation of state and conductivity of a collisionless plasma, *Phys. Fluids* **4**, 669 (1961).

448. Burkhardt, G., Fahl, C., and Larenz, R., Coupling mechanism between longitudinal and transverse waves in a plasma, *Z. Physik* **161**, 380 (1961).

449. Buti, B., Plasma oscillations and Landau damping in a relativistic gas, *Phys. Fluids* **5**, 1 (1962).
450. Callen, H. and Green, R., On the thermodynamics of irreversible processes, *Phys. Rev.* **86**, 702 (1952).
451. Callen, H. and Welton, T., Irreversibility and generalized noise, *Phys. Rev.* **83**, 34 (1951).
452. Canobbio, E. and Croci, R., Modulated beams in a plasma in the presence of a magnetic field, *Z. Naturforsch.* **16a**, 1313 (1961).
453. Capps, R. H., Properties of intense relativistic electron beams, *Phys. Rev.* **114**, 1203, 1959.
454. Case, K., Plasma oscillations, *Ann. Phys.* **7**, 349 (1959).
455. Chandrasekhar, S., Stochastic problems in physics and astronomy, *Rev. Mod. Phys.* **15**, 1 (1943).
456. Chapman, S., Thermodiffusion in ionized gases, *Proc. Phys. Soc. (London)* **72**, 353 (1958).
457. Chapman, S. and Cowling, T., "The Mathematical Theory of Non-Uniform Gases," Cambridge Univ. Press, London and New York, 1952.
458. Chau-Hsing, Su, Variational principles in plasma dynamics, *Phys. Fluids* **4**, 1376 (1961).
459. Chew, G., Goldberger, M., and Low, F., Boltzmann equation and a one-fluid hydromagnetic equation in the absence of collisions between particles, *Proc. Roy. Soc. (London)* **A236**, 112 (1956).
460. Clemmow, P. C. and Willson, A. J., Dispersion equation for plasma oscillations, *Proc. Roy. Soc. (London)* **A237**, 117 (1956).
461. Clemmow, P. C. and Willson, A. J., Relativistic Boltzmann transport equations in the absence of collisions, *Proc. Cambridge Phil. Soc.* **53**, 222 (1958).
462. Cohen, M. H., Radiation in plasma. Cerenkov effect, *Phys. Rev.* **123**, 711 (1961).
463. Cohen, R., Spitzer, L., and Routly, P., Electrical conductivity of an ionized gas, *Phys. Rev.* **80**, 230 (1950).
464. Colgate, S. A., Shock waves in a collisionless plasma, *Phys. Fluids* **2**, 485 (1959).
465. Cowling, T. G., "Magnetohydrodynamics." Wiley (Interscience), New York, 1957.
466. Czyz, W. and Gottfried, K., Distribution of momenta in a Fermi gas at zero temperature, *Nucl. Phys.* **21**, 676 (1961).
467. Darwin, C., Coefficient of refraction of an ionized medium, *Proc. Roy. Soc. (London)* **146**, 17 (1934); **182**, 152 (1944).
468. Dawson, J. M., Nonlinear electron oscillations in a cold plasma, *Phys. Rev.* **113**, 383 (1959).
469. Dawson, J. M., Plasma oscillations in the presence of a large number of electron beams, *Phys. Rev.* **118**, 381 (1960).
470. Dawson, J. M., Landau damping, *Phys. Fluids* **4**, 869 (1961).
471. Dawson, J. M., One-dimensional model for a plasma, *Phys. Fluids* **5**, 445 (1962).
472. Dawson, J. and Oberman, C., Oscillations in a bounded cold plasma in a strong magnetic field, *Phys. Fluids* **2**, 103 (1959).
473. Dawson, J. and Oberman, C., High-frequency conductivity and coefficients

of emission and absorption of a fully ionized plasma, *Phys. Fluids* **5**, 517 (1962).

474. Desloges, E. A., Matthysse, S. W., and Margenau, H., Plasma conductivity in the microwave domain, *Phys. Rev.* **112**, 1437 (1958).

475. Dougal, A. A. and Goldstein, L., Energy exchange between electrons and ions in a plasma because of Coulomb collisions, *Phys. Rev.* **109**, 615 (1958).

476. Dougherty, J. P. and Farley, D. T., Theory of noncoherent radiowave scattering in a plasma, *Proc. Roy. Soc. (London)* **A259**, 79 (1960).

477. Doyle, P. H. and Neufeld, J., On the behavior of a plasma near ion resonance, *Phys. Fluids* **2**, 390 (1959).

478. Dreicer, H., Electron and ion escape in a fully ionized gas, *Phys. Rev.* **115**, 238 (1959); **117**, 329 (1960).

479. Dreicer, H., Electron velocity distribution in a partially ionized gas, *Phys. Rev.* **117**, 343 (1960).

480. Drummond, J. E., Microwave propagation in a hot magnetoactive plasma, *Phys. Rev.* **112**, 1460 (1958).

481. Drummond, J. E., "Plasma Physics." McGraw-Hill, New York, 1961.

482. Drummond, J. E., Cyclotron radiation of a hot plasma, *J. Nucl. Energy* **C2**, 90 (1961).

483. Drummond, J. E., Gerwin, R. A., and Springer, B. G., Conductivity concept, *J. Nucl. Energy* **C2**, 98 (1961).

484. Drummond, J. E. and Rosenbluth, M. N., Cyclotron radiation in a hot plasma, *Phys. Fluids* **3**, 45 (1960).

485. Du Bois, D., Theory of the electron field; degenerate gas, *Ann. Phys.* **7**, 174 (1959).

486. Du Bois, D., Properties of a dense electron gas, *Ann. Phys.* **8**, 24 (1959).

487. Du Bois, D., Gilinsky, V., and Kivelson, M., Attenuation of plasma oscillations because of collisions, *Phys. Rev. Letters* **8**, 419 (1962).

488. Dungly, J. W., Strong hydromagnetic excitations in a collisionless plasma, *Phil. Mag.* **4**, 585 (1959).

489. Dupree, T. H., Dynamics of an ionized gas, *Phys. Fluids* **4**, 696 (1961).

490. Edwards, S. F., New method of calculating the electrical conductivity of metals, *Phil. Mag.* **3**, 1020 (1958).

491. Edwards, S. F., Correlation function for charge in a plasma in the presence of a magnetic field, *Phil. Mag.* **6**, 61 (1961).

492. Edwards, S. F. and Sanderson, J. J., New approach to transport problems in a fully ionized plasma, *Phil. Mag.* **6**, 71 (1961).

493. Ehrenreich, H. and Cohen, M., Self-consistent field approximation to the multi-electron problem, *Phys. Rev.* **115**, 786 (1959).

494. Elsasser, W., Theory of the hydromagnetic dynamo, *Rev. Mod. Phys.* **28**, 135 (1956).

495. Engelhardt, A. G. and Dougal, A. A., Dispersion of ionic cyclotron waves in a magnetoplasma, *Phys. Fluids* **5**, 29 (1962).

496. Engelsberg, S., Energy losses in a many particle system, *Phys. Rev.* **123**, 1130 (1961).

497. Englert, F. and Brout, R., Dielectric description of the quantum statistics of interacting particles, *Phys. Rev.* **120**, 1085 (1960).

498. Enoch, J., Nonlinear theory of transverse plasma oscillations, *Phys. Fluids* **5**, 467 (1962).

499. Epstein, P. S., Theory of wave propagation in a hydromagnetic medium, *Rev. Mod. Phys.* **28**, 3 (1956).

500. Fang, P. H., Plasma conductivity in the microwave domain, *Phys. Rev.* **113**, 13 (1959).

501. Fano, U., Atomic theory of electromagnetic interactions in dense media, *Phys. Rev.* **103**, 1202 (1956).

502. Farley, D. T., Dougherty, J. P., and Barron, D. W., Theory of noncoherent radiowave scattering in a plasma. 2. Scattering in a magnetic field, *Proc. Roy. Soc. (London)* **A263**, 238 (1961).

503. Fejer, J., Radiowave scattering by an ionized gas in thermal equilibrium, *Canad. J. Phys.* **38**, 1115 (1960).

504. Fejer, J. A., Radiowave scattering in an ionized gas in thermal equilibrium in the presence of a homogeneous magnetic field, *Canad. J. Phys.* **39**, 716 (1961).

505. Fermi, E., Ionization energy losses in gases and dense media, *Phys. Rev.* **57**, 485 (1940).

506. Fermi, E. and Teller, E., Absorption of negative mesons in a substance, *Phys. Rev.* **72**, 399 (1947).

507. Ferraro, V., Plasma physics and magnetohydrodynamics. General theory of plasma. *Nuovo Cimento Suppl.* **13**, 9 (1959).

508. Ferrell, R. A., Dispersion relation and least wavelength for plasma oscillations, *Phys. Rev.* **107**, 450 (1957).

509. Ferrell, R. A. and Quinn, J. J., Characteristic energy losses of electrons passing through metal films. Exciton model of plasma oscillations, *Phys. Rev.* **108**, 570 (1957).

510. Ford, G., Electromagnetic radiation in a plasma, *Ann. Phys.* **16**, 185 (1961).

511. Fowler, T. K., Plasma stability relative to electrostatic perturbations, *Phys. Fluids* **4**, 1393 (1961).

512. Fried, B. D. and Gould, R. W., Longitudinal ion oscillations in a hot plasma, *Phys. Fluids* **4**, 139 (1961).

513. Fried, B. D. and Conte, S. D., "The Plasma Dispersion Function." Academic Press, New York, 1961.

514. Frisch, H. L., Thermalization time of fast ions in a plasma, *Phys. Fluids* **4**, 1167 (1961).

515. Frohlich, H. and Pelzer, H., Plasma oscillations and energy losses of charged particles in solids, *Proc. Phys. Soc. (London)* **A68**, 525 (1955).

516. Fujita, E. and Usui, T., Excitations in a dense electron gas. 2. Diamagnetism, *Progr. Theor. Phys.* **23**, 799 (1961).

517. Gabor, D., Plasma oscillations, *Trans. IRE* **Ap-4**, 526 (1956).

518. Gambirasio, G., Electrical properties of an ideal plasma, *Phys. Fluids* **3**, 299 (1960).

519. Gartenhaus, S., On a variational principle for a classical plasma, *Phys. Fluids* **4**, 1122 (1961).

520. Gasiorowicz, S., Neumann, M., and Ridell, R., Dynamics of an ionized medium, *Phys. Rev.* **101**, 922 (1956).

521. Gell-Mann, M. and Brueckner, K. A., Correlation energy of a high-density electron gas, *Phys. Rev.* **106**, 364 (1957).

522. Glassgold, A. E., Collective oscillations of a system of interacting Fermions, *J. Nucl. Energy* **C2**, 51 (1961).

523. Glick, A. J., Linear influence function of a system of Fermi-particles, *Ann. Phys.* **17**, 61 (1962).

524. Glick, A. J. and Ferrell, R. A., Single-particle excitations in a degenerate electron gas, *Ann. Phys.* **11**, 359 (1960).

525. Goldstein, L. and Sekiguchi, T., Electron-electron interaction and plasma heat conduction, *Phys. Rev.* **109**, 625 (1958).

526. Gordon, W., Noncoherent radiowave scattering by free electrons, *Proc. I. R. E.* **46**, 1824 (1958).

527. Goto, K., Characteristic functional for turbulence in a plasma, *Progr. Theor. Phys.* **25**, 603 (1961).

528. Gottfried, K. and Picman, L., Sound propagation in a rarefied Fermi-gas at zero temperature, *Dan. Mat. Fys. Medd.* **32**, No. 13 (1960).

529. Green, H. S., Theory of an ionic plasma and magnetohydrodynamics, *Phys. Fluids* **2**, 341 (1959).

530. Greenwood, D. A., Kinetic equation in the theory of electrical conductivity of metals, *Proc. Phys. Soc. (London)* **71**, 585 (1958).

531. Greifinger, P. S., Excitation of oscillations in a rarefied plasma in the presence of a magnetic field, *Phys. Fluids.* **4**, 104 (1961).

532. De Groot, S. R., Thermodynamically nonequilibrium systems in an electromagnetic field, *J. Nucl. Energy* **C2**, 188 (1961).

533. Gross, E. P., Plasma oscillations in a permanent magnetic field, *Phys. Rev.* **82**, 232 (1951).

534. Gross, E. P., Dynamics of an electron stream and plasma, *Proc. Symp. Electronic Waveguides*, New York, 1958.

535. Gross, E. P., Dynamics of classical multi-particle systems, *J. Nucl. Energy* **C2**, 173 (1961).

536. Gross, E.P. and Krook, M., Model for collision processes in gases. Low amplitude charge oscillations in two-component systems, *Phys. Rev.* **102**, 593 (1956).

537. Hagfors, Tor., Density fluctuations in a plasma in a magnetic field and application to the ionosphere, *J. Geophys. Res.* **66**, 1699 (1961).

538. Harris, E. G., Instability of a plasma associated with the anisotropic velocity distribution, *J. Nucl. Energy* **C2**, 138 (1961).

539. Harris, E. G. and Simon, A., Coherent and noncoherent radiation in a plasma, *Phys. Fluids* **3**, 255 (1960).

540. Harris, G. M. and Trulio, J., Equilibrium properties of a partially ionized plasma, *J. Nucl. Energy* **C2**, 224 (1961).

541. Harrison, E. R., Effect of electron escape in a fully ionized plasma, *Phil. Mag.* **3**, 1318 (1958).

542. Hayakawa, S. and Hokkyo, N., Electromagnetic radiation from an electron plasma, *Progr. Theor. Phys.* **15**, 193 (1956).

543. Hayakawa, S. and Kitao, K., Energy losses of a charged particle passing through an ionized gas, and cosmic ray injection energy, *Progr. Theor. Phys.* **16**, 139 (1956).

544. Hayes, J. N., Attenuation of plasma oscillations in linear theory, *Phys. Fluids* **4**, 1387 (1961).

545. Hernquist, K. G., Plasma oscillations in electron beams, *J. Appl. Phys.* **26**, 1029 (1955).

546. Hide, R., Waves in a heavy, viscous, incompressible conducting fluid of variable density in the presence of a magnetic field, *Proc. Roy. Soc. (London)* **A233**, 376 (1955).

547. Hirshfield, J. L., Baldwin, D. E., and Brown, S. C., Cyclotron radiation in a hot plasma, *Phys. Fluids* **4**, 198 (1961).

548. Hirshfield, J. L. and Brown, S. C., Noncoherent microwave radiation in a plasma in a magnetic field, *Phys. Rev.* **122**, 719 (1961).

549. Hubbard, J., On electron interaction in metals, *Proc. Phys. Soc. (London)* **A62**, 441, 977 (1955).

550. Hubbard, J., Dielectric description of electron interaction in solids, *Proc. Phys. Soc. (London)* **A68**, 976 (1955).

551. Hubbard, J., Description of collective motion in terms of many-particle perturbation theory, *Proc. Roy. Soc. (London)*, **A240**, 539 (1957).

552. Hubbard, J., Correlation energy of a free-electron gas, *Proc. Roy. Soc. (London)*, **A243**, 337, 1958.

553. Hubbard, J., Generalization of the theory to the case of an inhomogeneous gas, *Proc. Roy. Soc. (London)* **A244**, 199 (1958).

554. Hubbard, J., Friction and diffusion coefficients for the Fokker-Planck equation in a plasma. 1, *Proc. Roy. Soc. (London)* **A260**, 114 (1961).

555. Hubbard, J., Friction and diffusion coefficients for the Fokker-Planck equation in a plasma, *Proc. Roy. Soc. (London)* **A261**, 371 (1961).

556. Hudson, J., Landau damping for non-Maxwellian distributions, *Proc. Cambridge Phil. Soc.* **58**, 119 (1962).

557. Hwa, R. C., Influence of electron–electron interaction on cyclotron resonance in a gas plasma, *Phys. Rev.* **110**, 307 (1958).

558. Ichikawa, Y. H., Retarding force of a high-temperature plasma. Effects of collective ion motion, *Progr. Theor. Phys.* **23**, 512 (1960).

559. Ichikawa, Y. H., On the kinetic equation for a high-temperature plasma, *Progr. Theor. Phys.* **24**, 1083 (1960).

560. Ichikawa, Y. H. and Sasakura, Y., On the structure of the generalized Fokker-Planck equation for a high-temperature plasma, *Progr. Theor. Phys.* **25**, 989 (1961).

561. Ichikawa, Y. H. and Yamamoto, M., Bremsstrahlung in a dense medium at high energies, *Progr. Theor. Phys.* **23**, 81 (1960).

562. Imre, K., Oscillations of a relativistic plasma, *Phys. Fluids* **5**, 459 (1962).

563. Izuyama, T., Electrical conductivity for a longitudinal electric field, *Progr. Theor. Phys.* **25**, 964 (1961).

564. Jackson, E. A., Drift instabilities in a Maxwellian plasma, *Phys. Fluids* **3**, 786 (1960).

565. Jackson, E. A., Nonlinear oscillations in a cold plasma, *Phys. Fluids* **3**, 831 (1960).

566. Jelley, J., "Cerenkov Radiation". Pergamon Press, New York, 1958.

567. Johnson, P. S., Cerenkov radiation spectrum in a cold magnetoactive plasma, *Phys. Fluids* **5**, 118 (1962).
568. Jukes, J. D., High-frequency portion of cyclotron radiation in a hot plasma, *Phys. Fluids* **4**, 1184 (1961).
569. Kaji J., and Kito, M., Plasma oscillations in a magnetic field, *J. Phys. Soc. (Japan)* **15**, 1851 (1960).
570. Kalman, G. and Ron, A., Energy losses of an experimental particle, *Ann. Phys.* **16**, 118 (1961).
571. Kanazawa, H., On plasma oscillations in metal films, *Progr. Theor. Phys.* **26**, 851 (1961).
572. Kanazawa, H. and Watabe, M., Green's function method for an electron gas. 1. General formulation, *Progr. Theor. Phys.* **23**, 408 (1960).
573. Kanazawa, H., Misawa, S., and Fujita, E., Green's function method for an electron gas. 2. Dispersion relations for plasmons, *Progr. Theor. Phys.* **23**, 426 (1960).
574. Kanazawa, H. and Matsudaira, N., Green's function method for an electron gas. 3. Diamagnetism, *Progr. Theor. Phys.* **23**, 433 (1960).
575. Karplus, R., Radiation of hydromagnetic waves, *Phys. Fluids* **3**, 800 (1960).
576. Kellogg, P. I. and Ziemolh, H., Instability of mutually penetrating plasmas, *Phys. Fluids* **3**, 40 (1960).
577. Kelly, D. C., Microwave conductivity of a plasma in a magnetic field, *Phys. Rev.* **119**, 27 (1960).
578. Kidder, R. E. and de Witt, H. E., Application of a modified Debye-Hueckel theory to fully ionized gases, *J. Nucl. Energy* **C2**, 218 (1961).
579. Kihara, T., Relaxation between ions and electrons in a plasma in a strong magnetic field, *J. Phys. Soc. (Japan)* **14**, 1751 (1959); **15**, 684 (1960).
580. Kihara, T., Aono, O., and Sugihara, R., Theory of Cerenkov and cyclotron resonance in plasma, *Nuclear Fusion* **1**, 181 (1961).
581. Kildal, A., Landau damping and Gross gaps in the frequency spectrum of plasma oscillations, *Nuovo Cimento* **20**, 104 (1961).
582. Kischel, K., Theory of electric waves in an inhomogeneous plasma, *Ann. Physik* **B19**, 309 (1957).
583. Kitao, K., Energy losses and radiation of a rotating charged particle in a magnetic field, *Progr. Theor. Phys.* **23**, 759 (1960).
584. Koga, T., Transport phenomena in a fully ionized gas in the presence of a strong magnetic field, *Phys. Fluids* **4**, 834 (1961).
585. Koga, T., Interaction between radiowaves and a plasma, *Phys. Fluids* **4**, 1162 (1961).
586. Koga, T., Kinetic equations for a plasma, *Phys. Fluids* **5**, 705 (1962).
587. Koga, T., Everton, J., and Wilber, P., Model of electron collisions with heavy particles in a plasma, *Phys. Fluids* **4**, 1057 (1961).
588. Kolb, A. C., Magnetic compression of a plasma, *Rev. Mod. Phys.* **32**, 748 (1960).
589. Körper, K., Motion of a plasma cylinder in an external magnetic field, *Z. Naturforsch.* 12a, 815 (1957).
590. Krall, N. and Rosenbluth, M., Stability of a weakly inhomogeneous plasma, *Phys. Fluids* **4**, 163 (1961).

591. Kramers, H. A., Retarding capacity of metals for alpha-particles, *Physica* **13**, 401 (1947).

592. Kranzer, H. C., Thermalization of fast ions in a plasma, *Phys. Fluids* **4**, 214 (1961).

593. Kraus, L. and Watson, K. M., Plasma motions due to satellites in the ionosphere, *Phys. Fluids* **1**, 480 (1959).

594. Kritz, A. H. and Mintzer, D., Plasma wave propagation in the presence of density jumps, *Phys. Rev.* **117**, 382 (1960).

595. Kronig, R., Remark on the retardation of fast charged particles in metal conductors, *Physica* **15**, 667 (1949).

596. Kronig, R. and Korringa, J., On the theory of deceleration of fast charged particles in metals, *Physica* **10**, 406 (1943).

597. Kubo, R., Statistical mechanics of irreversible processes. 1. General theory and some simple applications to magnetism and electrical conduction problems, *J. Phys. Soc. (Japan)* **12**, 570 (1957).

598. Kubo, R., "Lectures in Theoretical Physics," Vol. 1. Wiley (Interscience), New York, 1959.

599. Kubo, R., Yokota, M., and Nakajima, S., Statistical mechanics of irreversible processes. 2. Reactions to thermal excitation, *J. Phys. Soc. (Japan)* **12**, 1203 (1957).

600. Kursunoglu, B., Relativistic plasma, *Nuclear Fusion* **1**, 213 (1961).

601. Laaspere, T., Electromagnetic wave scattering in the ionosphere, *J. Geophys. Res.* **65**, 3955 (1960).

602. Landau, L. D., Diamagnetism of metals, *Z. Physik* **64**, 629 (1930).

603. Langmuir, I., Oscillations in ionized gases, *Proc. Natl. Acad. Sci.* **14**, 627 (1926).

604. Lax, M., Generalized mobility theory, *Phys. Rev.* **109**, 1921 (1958).

605. Lax, M., Fluctuations in a nonequilibrium stable state, *Rev. Mod. Phys.* **32**, 25 (1960).

606. Lehnert, B., Plasma of cosmic and laboratory scales, *Nuovo Cimento Suppl.* **13**, 59 (1959).

607. Lenard, A., On the Bogoliubov kinetic equation for a spatially homogeneous plasma, *Ann. Phys.* **3**, 390 (1960).

608. Lenard, A. and Bernstein, I. B., Plasma oscillations and diffusion in velocity space, *Phys. Rev.* **112**, 1456 (1958).

609. Levine, H. B., Diagram classification for the plasma equation of state, *J. Nucl. Energy* **C2**, 206 (1961).

610. Lindhard, J., On properties of a gas of charged particles, *Dan. Mat. Fys. Medd.* 28, No. 8 (1954).

611. Linhart, J. G., "Plasma Physics." North-Holland, Amsterdam, 1961.

612. Linhart, J. and Schoch, A., Thermonuclear apparatus with a relativistic electron beam, *Nucl. Instr. Methods* **4**, 332 (1959).

613. Longmire, C. and Rosenbluth, M., Diffusion of charged particles in a magnetic field, *Phys. Rev.* **103**, 507 (1956).

614. Loughhead, R., Natural oscillations of a compressed ionized medium, *Austral. J. Phys.* **8**, 416 (1955).

615. Lukes, T., Fluctuations and electrical conductivity of metals, *Physica* **27**, 319 (1961).
616. Lundquist, S., Escape phenomena and relaxation effects, *Rend. Scuola Internaz. Fis. "Enrico Fermi"* **13** (1959).
617. MacDonald, W., Rosenbluth, M., and Judd, D., Fokker-Planck equation for forces inversely proportional to the distance squared, *Phys. Rev.* **107**, 1 (1957).
618. MacDonald, W. M., Rosenbluth, M. N., and Wong, C., Relaxation in a system of particles with Coulomb interaction, *Phys. Rev.* **107**, 350 (1957).
619. Majumdar, S., Radiation by charged particles passing through an electron plasma in an external magnetic field, *Proc. Phys. Soc. (London)* **77**, 1109 (1961).
620. Mannari, I., Space-time correlation function in the theory of electrical conductivity, *Progr. Theor. Phys.* **26**, 51 (1961).
621. Margenau, H., Plasma conductivity in the microwave domain, *Phys. Rev.* **100**, 6 (1958).
622. Marshall, W., Structure of magnetohydrodynamic shockwaves, *Proc. Roy. Soc. (London)* **A233**, 367 (1955).
623. Marshall, W., Structure of magnetohydrodynamic shockwaves in a plasma with infinite conductivity, *Phys. Rev.* **103**, 1900 (1956).
624. Martin, P. C. and Schwinger, J., Theory of systems of many particles, *Phys. Rev.* **115**, 1342 (1959).
625. Marx, G. and Gyorgyi, G., On the energy-momentum tensor for an electromagnetic field in a dielectric, *Ann. Physik* **16**, 241 (1955).
626. Matsubara, T., New method in quantum statistics, *Progr. Theor. Phys.* **14**, 351 (1955).
627. Medicus, G., Diffusion and losses because of elastic collisions of fast electrons in a plasma, *J. Appl. Phys.* **29**, 903 (1958).
628. Meecham, W. C., Problem of the source and reflection in a magnetoionic medium, *Phys. Fluids* **4**, 1517 (1961).
629. Meeron, E., Transport equations for plasma in strong external fields, *Phys. Rev.* **124**, 308 (1961).
630. Mitchner M., ed. "Radiation and Waves in Plasma." Stanford Univ. Press, Stanford, California, 1961.
631. Montgomery, D., Time-dependent nonlinear plasma oscillations, *Phys. Rev.* **123**, 1077 (1961).
632. Montgomery, D. and Gorman, D., Landau damping in high approximations, *Phys. Rev.* **124**, 1309 (1961).
633. Montroll, E. and Ward, J., Quantum statistics of interacting particles. 1. General theory and some remarks on the properties of an electron gas, *Phys. Fluids* **1**, 55 (1958).
634. Montroll, E. and Ward, J., Quantum statistics of interacting particles. 2. Group decomposition of the kinetic coefficients, *Physica* **25**, 423 (1959).
635. Mori, H., Quantum-statistical theory of transport processes, *J. Phys. Soc. (Japan)* **11**, 1029 (1956).
636. Mower, L., Conductivity of a hot plasma, *Phys. Rev.* **116**, 16 (1959).

637. Nakajima, S., On the electromagnetic influence function for normal metals, *Progr. Theor. Phys.* **23**, 694 (1960).

638. Nakano, H., Method of calculating the electrical conductivity, *Progr. Theor. Phys.* **15**, 77 (1954); **17**, 145 (1957).

639. Neufeld, J., Vavilov–Cerenkov radiation in a high-temperature plasma, *Phys. Rev.* **116**, 1 (1959).

640. Neufeld, J., Electromagnetic wave propagation in a many-beam medium near gyromagnetic resonance, *Phys. Rev.* **116**, 19 (1959).

641. Neufeld, J. and Doyle, P. H., Electromagnetic interaction of a charged-particle beam with a plasma, *Phys. Rev.* **121**, 654 (1961).

642. Neufeld, J., Properties of spatial dispersion in a plasma, *Phys. Rev.* **123**, 1 (1961).

643. Neufeld, J. and Ritchie, R. H., Passage of charged particles through plasma, *Phys. Rev.* **98**, 1632 (1955).

644. Neufeld, J. and Wright, H., Vavilov–Cerenkov effect and Bohr radiation due to a beam of charged particles in a dispersive medium, *Phys. Rev.* **124**, 1 (1961).

645. Nexsen, W. E., Cummins, Jr. W., Coensgen, F. H., and Sherman, A. E., Collision of two plasma beams, *Phys. Rev.* **119**, 1457 (1960).

646. Noerdlinger, P. D., Stability of a homogeneous plasma relative to longitudinal oscillations, *Phys. Rev.* **118**, 879 (1960).

647. Northop, T. G., Helmholtz instability in plasma, *Phys. Rev.* **103**, 1150 (1956).

648. Nozieres, P. and Pines, D., Dielectric description of many-body problems. Free electron gas, *Nuovo Cimento* **9**, 470 (1958).

649. Nozieres, P. and Pines, D., Electron interaction in solids, *Phys. Rev.* **109**, 741 (1958).

650. Nozieres, P. and Pines, D., Collective description of the dielectric constant, *Phys. Rev.* **109**, 762 (1958).

651. Nozieres, P. and Pines, D., Electron interaction in solids. Spectrum of characteristic energy losses, *Phys. Rev.* **113**, 1254 (1959).

652. Nyquist, H., Thermal fluctuations of an electric charge in conductors, *Phys. Rev.* **32**, 110 (1928).

653. Oberman, C., Radiation due to collective phenomena in a plasma, *J. Nucl. Energy* **C2**, 154 (1961).

654. Oster, L., Spectral and angular distribution of cyclotron radiation emitted during particle collisions, *Phys. Rev.* **116**, 474 (1959).

655. Oster, L., Linear theory of plasma oscillations, *Rev. Mod. Phys.* **32**, 141 (1960).

656. Oster, L., Cyclotron radiation of relativistic particles with arbitrary velocity distribution, *Phys. Rev.* **121**, 961 (1961).

657. Pai, S. I., Low-amplitude wave propagation in a fully ionized plasma in the presence of a magnetic field, *Phys. Fluids* **5**, 234 (1962).

658. Pappert, R., Excitation of plasma oscillations by a charge in a magnetic field, *Phys. Fluids* **3**, 966 (1960).

659. Parker, E., Hydromagnetic waves and acceleration of cosmic rays, *Phys. Rev.* **99**, 241 (1955).

660. Parker, E. N., Quasi-linear model of a plasma shock structure in a longitudinal magnetic field, *J. Nucl. Energy* **C2**, 146 (1961).

Bibliography   239

661. Penrose, O., Electrostatic instability in a homogeneous non-Maxwellian plasma, *Phys. Fluids* **3**, 258 (1960).
662. Piddington, J., Hydrodynamic waves in an ionized gas, *Phil. Mag.* **46**, 1037 (1955).
663. Pierce, J. R., Critical remarks on the theory of devices with electron beams, *J. Nucl. Energy* **C2**, 73 (1961).
664. Pines, D., Energy losses of charged particles in metals, *Phys. Rev.* **85**, 931 (1952).
665. Pines, D., Collective description of electron interaction. 4. Electron interaction in metals, *Phys. Rev.* **92**, 626 (1953).
666. Pines, D., Collective energy losses in solids, *Rev. Mod. Phys.* **28**, 184 (1956).
667. Pines, D., Plasma oscillations in an electron gas, *Physica* **26**, 103 (1960).
668. Pines, D., Classical and quantum plasma, *J. Nucl. Energy* **C2**, 5 (1961).
669. Pines, D. and Bohm, D., Collective description of electron interaction. 2. Coulomb interaction, *Phys. Rev.* **85**, 338 (1952).
670. Pines, D. and Schrieffer, J. R., Buildup of equilibrium between electrons, plasmons, and phonons in a plasma in the classical and quantum cases, *Phys. Rev.* **125**, 804 (1962).
671. Platzman, P. M. and Buchsbaum, S. J., Influence of collisions on Landau damping of plasma oscillations, *Phys. Fluids* **4**, 1288 (1961).
672. Poeverlein, H., Electromagnetic wave propagation in a plasma in the presence of a strong magnetic field, *Phys. Fluids* **4**, 397 (1961).
673. Post, R., Investigations on controlled synthesis. Applications of high-temperature plasma physics, *Rev. Mod. Phys.* **28**, 338 (1956).
674. Post, R. F., High-temperature plasma and controlled thermonuclear reactions, *Ann. Rev. Nucl. Sci.* **9**, 367 (1959).
675. Pradhan, T., Plasma oscillations in a permanent magnetic field, *Phys. Rev.* **107**, 1222 (1957).
676. Pradhan, T., Causality and dispersion relations for waves in a plasma, *Ann. Phys.* **17**, 418 (1962).
677. Prigogine, I., On charged particle motion, *J. Nucl. Energy* **C2**, 184 (1961).
678. Quinn, J. J. and Ferrell, R. A., Correlation energy of a degenerate electron gas, *Phys. Rev.* **112**, 812 (1958).
679. Quinn, J. J. and Ferrell, R. A., Quasi-particle description of interaction in an ideal metal, *J. Nucl. Energy* **C2**, 18 (1961).
680. Rand, S., Electrostatic field of a slowly moving ion in a plasma, *Phys. Fluids* **2**, 649 (1959).
681. Rand, S., Deceleration of a satellite moving in the ionosphere, *Phys. Fluids* **3**, 265, 588 (1960).
682. Rand, S., Plasma as a microwave converter, *Phys. Fluids* **4**, 860 (1961).
683. Rand, S., Coefficient of dynamic friction for slow ions, *Phys. Fluids* **4**, 1251 (1961).
684. Ratcliffe, J. A., "The Magnetoionic Theory and Its Applications to the Ionosphere," Cambridge University Press, London and New York, 1959.
685. Rawer, K. and Suchy, K., Longitudinal and transverse waves in a Lorentz plasma, *Ann. Physik* **7**, 155 (1959).

686. Rawer, K. and Suchy, K., Statistical deviations in the dispersion formula for a Lorentz plasma at finite temperature, *Ann. Physik* **7**, 313 (1958).

687. Renau, J., Electromagnetic wave scattering in the ionosphere, *J. Geophys. Res.* **65**, 3631 (1960).

688. Ritchie, R. H., Collective energy losses of electrons in films, *Phys. Rev.* **106**, 874 (1957).

689. Ritchie, R. H., Charged-particle interaction with a degenerate electron gas, *Phys. Rev.* **114**, 644 (1959).

690. Robinson, B. B. and Bernstein, I. B., Variational description of transport processes in a plasma, *Ann. Phys.* **18**, 110 (1962).

691. Rockmore, R. M., Energies of single-particle excitations in a high-density degenerate electron gas, *Phys. Rev.* **114**, 941 (1959).

692. Ron, A., Penetration of a magnetic field into a plasma, *Nuovo Cimento* **10**, 659 (1958).

693. Ron, A. and Kalman, G., Interaction of a test particle with a plasma. Distribution function of field particles, *Ann. Phys.* **11**, 240 (1960).

694. Ron, A. and Kalman, G., Correlations in a statistical distribution of electrical microfields in a plasma dependent on the velocity, *Phys. Rev.* **123**, 1100 (1961).

695. von Roos, O., Boltzmann-Vlasov equation for a quantum plasma, *Phys. Rev.* **119**, 1174 (1960).

696. Rosen, P., Electromagnetic wave scattering by longitudinal plasma waves, *Phys. Fluids* **3**, 416 (1960).

697. Rosenbluth, M. N. and Longmire, C. L., Stability of a plasma in a magnetic field, *Ann. Phys.* **1**, 120 (1957).

698. Rosenbluth, M. N. and Rostoker, N., Structure of the equations of a plasma, *Phys. Fluids* **2**, 23 (1959).

699. Rosenbluth, M. N. and Rostoker, N., Electromagnetic wave scattering in a nonequilibrium plasma, *Phys. Fluids* **5**, 776 (1962).

700. Rostoker, N., Fluctuations in plasma, *Nuclear Fusion* **1**, 101 (1961).

701. Rostoker, N. and Rosenbluth, M. N., Fokker-Planck equation for a plasma in a permanent magnetic field, *J. Nucl. Energy* **C2**, 195 (1961).

702. Salpeter, E. E., Fluctuations of electron density in a plasma, *Phys. Rev.* **120**, 1528 (1960).

703. Salpeter, E. E., Radiowave scattering in the ionosphere, *J. Geophys. Res.* **65**, 1851 (1960); **66**, 982 (1961).

704. Salpeter, E. E., Density fluctuations of a plasma in a magnetic field, *Phys. Rev.* **122**, 1663 (1961).

705. Sawada, K., Correlation energy of a high density electron gas, *Phys. Rev.* **106**, 372 (1957).

706. Sawada, K., Brueckner, K., Fukuda, N., and Brout, R., Correlation energy of a high-density electron gas. Plasma oscillations, *Phys. Rev.* **108**, 507 (1957).

707. Scarf, F. L., Landau damping and absorption of low-frequency signals, *Phys. Fluids* **5**, 6 (1962).

708. Seizo, U., Dual correlation functions of a nonideal electron gas at high densities, *Progr. Theor. Phys.* **26**, 45 (1961).

709. Sen, H., Magnetohydrodynamic shockwave structure in an ideally conducting plasma, *Phys. Rev.* **102**, 5 (1956).

710. Schirmer, J. and Friedrich H., Plasma heat conductivity, *Z. Physik* **153**, 563 (1959).

711. Schlüter, A., Gyrorelaxational effect, *Z. Naturforsch.* **12a**, 822 (1957).

712. Schwartzschild, M. and Kruskal, M., On instability of a fully ionized plasma, *Proc. Roy. Soc. (London)* **123**, 348 (1954).

713. Sehoguchi, J. and Herncha, R. C., Heat conductivity of an electron gas in a plasma, *Phys. Rev.* **112**, 1 (1958).

714. Simon, A., Relativistic Fokker-Planck coefficients for a plasma and radiation, *Phys. Fluids* **4**, 691 (1961).

715. Simon, A. and Harris, E. G., Kinetic equations for a plasma and radiation, *Phys. Fluids* **3**, 245 (1960).

716. Spitzer, L., Transport phenomena in a fully ionized gas, *Phys. Rev.* **89**, 977 (1953).

717. Spitzer, L., "Physics of Fully Ionized Gases." 2nd ed. Wiley (Interscience), New York, 1962.

718. Spitzer, L., Collective phenomena in a hot plasma, *Nature* **181**, 221 (1958).

719. Spitzer, L. and Harm, R., Heat conductivity of an electron gas, *Phys. Rev.* **89**, 977 (1953).

720. Stephen, M. J., Boltzmann gas of charged particles at absolute zero, *Proc. Phys. Soc.* **79**, 994 (1962).

721. Stern, E. A. and Ferrell, R. A., Surface plasma oscillations of a degenerate electron gas, *Phys. Rev.* **120**, 130 (1960).

722. Stix, T., Oscillations of a cylindrical plasma, *Phys. Rev.* **106**, 1146 (1957).

723. Stix, T. H., Excitation and thermalization of plasma waves, *Phys. Fluids* **1**, 308 (1958).

724. Stix, T. H., Absorption of plasma waves, *Phys. Fluids* **3**, 19 (1960).

725. Stuetzer, O. M., Magnetohydrodynamics and electrodynamics, *Phys. Fluids* **5**, 534 (1962).

726. Sturrock, P. A., Nonlinear phenomena in an electron plasma, *Proc. Roy. Soc. (London)* **A242**, 277 (1957).

727. Sturrock, P. A., Kinematics of growing waves, *Phys. Rev.* **112**, 1488 (1958).

728. Sturrock, P. A., Variational principle and energy theorem for small perturbations of electron beams and electron–ion plasmas, *Ann. Phys.* **4**, 306 (1958).

729. Sturrock, P. A., Excitation of plasma oscillations, *Phys. Rev.* **117**, 1426 (1960).

730. Sturrock, P. A., Spectral characteristics of solar radio flares, *Nature* **192**, 58 (1961).

731. Sturrock, P. A., Nonlinear effects in an electron plasma, *J. Nucl. Energy* **C2**, 158 (1961).

732. Sumi, M., Theory of wave excitation in plasma, *J. Phys. Soc. (Japan)* **13**, 1476 (1958).

733. Sumi, M., Theory of plasma waves growing in space, *J. Phys. Soc. (Japan)* **14**, 653 (1959).

734. Takimoto, N., On shielding of magnetic interaction of degenerate free electrons, *Progr. Theor. Phys.* **24**, 923 (1960).

735. Tanenbaum, B. S., Dispersion relations in a stationary plasma, *Phys. Fluids* **4**, 1262 (1961).

736. Tang, C. L. and Meixner, J., Relativistic theory of plane electromagnetic wave propagation, *Phys. Fluids* **4**, 148 (1961).

737. Tayler, R. J., Influence of an axial magnetic field on the stability of a bounded gas discharge, *Proc. Phys. Soc.* **B70**, 1049 (1957).

738. Taylor, J. B., Correlations of an electric field and plasma dynamics, *Phys. Fluids* **3**, 792 (1960).

739. Taylor, J. B., Ion diffusion in a plasma across a magnetic field, *Phys. Fluids* **4**, 1142 (1961).

740. Tchen, C., Kinetic equation for plasma with nonsteady correlations, *Phys. Rev.* **114**, 394 (1959).

741. Ter Haar, D., "Introduction to the Physics of Many-Body Systems." Wiley (Interscience), New York and London, 1958.

742. Thompson, W. and Hubbard, J., Long-range forces and diffusion coefficients of a plasma, *Rev. Mod. Phys.* **32**, 714 (1960).

743. Tidman, D. A., Shockwave structure in fully ionized hydrogen, *Phys. Rev.* **111**, 1439 (1958).

744. Tidman, D. A., Radio radiation by plasma oscillations in an inhomogeneous plasma, *Phys. Rev.* **117**, 366 (1960).

745. Tidman, D. A., Radiation associated with plasma oscillations, *Phys. Fluids* **4**, 1186 (1961).

746. Tidman, D. A. and Boyd, J. M., Radiation by plasma oscillations passing through density discontinuities, *Phys. Fluids* **5**, 213 (1962).

747. Tidman, D. A. and Parker, E. N., Ultrathermal particles, *Phys. Rev.* **111**, 1206 (1958).

748. Tidman, D. A. and Weiss, G. H., Radio radiation by plasma oscillations in an inhomogeneous plasma, *Phys. Fluids* **4**, 703 (1961).

749. Tidman, D. A. and Weiss, G. H., Radiation by high-amplitude plasma oscillations, *Phys. Fluids* **4**, 866 (1961).

750. Tomonaga, S., On the applicability of the Bloch method to problems of many Fermi particles, *Progr. Theor. Phys.* **5**, 544 (1950).

751. Tonks, L. and Langmuir, I., Oscillations in ionized gases, *Phys. Rev.* **33**, 195, 990 (1929).

752. Trivelpiece, A. and Gould, R., Space charge waves in a cylindrical plasma column, *J. Appl. Phys.* **30**, 1784 (1959).

753. Trubnikov, B. A., Angular distribution of cyclotron radiation in a hot plasma, *Phys. Fluids* **4**, 195 (1961).

754. Twiss, R. Q., On the theory of circularly polarized wave amplification in an ionized medium, *Phys. Rev.* **84**, 448 (1951).

755. Ueda, S., Dual correlation function for a high-density electron gas, *Progr. Theor. Phys.* **26**, 45 (1961).

756. Usui, T., Excitations in a dense electron gas, *Progr. Theor. Phys.* **23**, 787 (1960).

757. Van Hove, L., Space-time correlations and scattering in a Born approximation for systems of interacting particles, *Phys. Rev.* **95**, 249 (1954).

758. Van Kampen, N. G., On the theory of stationary waves in a plasma, *Physica* **21**, 949 (1955).

759. Van Kampen, N. G., Dispersion equation for waves in a plasma, *Physica* **23**, 641 (1957).

760. Varma, R. K., Kinetic equations for plasma, *Phys. Fluids* **5**, 525 (1962).

761. Vernon, R., Influence of ion motion on nonlinear plasma oscillations, *Phys. Fluids* **4**, 1524 (1961).

762. Villars, F. and Weisskopf, V. F., Electromagnetic wave scattering by turbulent fluctuations in the atmosphere, *Phys. Rev.* **94**, 232 (1954).

763. Villars, F. and Weisskopf, V. F., Radiowave scattering by turbulent fluctuations in the atmosphere, *Proc. IRE* **43**, 1232 (1955).

764. Visvanathan, S., Magnetic properties of a relativistic electron gas, *Phys. Fluids* **5**, 701 (1962).

765. Warren, J. and Ferrell, R., Nonlocal coupling between the current and field in metals, *Phys. Rev.* **117**, 1252 (1960).

766. Watson, K. M., Use of the Boltzmann equation to study low-density ionized gases, *Phys. Rev.* **102**, 12 (1956).

767. Watson, K. M., Collective interactions in plasma, *Physica* **26**, 188 (1960).

768. Watson, K. M., Bludman, S. A., and Rosenbluth, M. N., Statistical mechanics of relativistic beams, *Phys. Fluids* **3**, 741; 747 (1960).

769. Wetzel, L., Anisotropy in an inhomogeneous plasma due to an electric field, *Phys. Rev.* **123**, 722 (1961).

770. Wilhelmsson, H., Stationary nonlinear plasma oscillations, *Phys. Fluids* **4**, 335 (1961).

771. Willis, C. R., Kinetic equation for the classical plasma, *Phys. Fluids* **5**, 219 (1962).

772. de Witt, H. E., Thermodynamic functions of a partially degenerate, fully ionized gas, *J. Nucl. Energy* **C2**, 27 (1961).

773. Wolff, P. A., Theory of plasma resonances, *Phys. Rev.* **103**, 845 (1956).

774. Yoshihara, H., Motion of small bodies in a rarefied plasma, *Phys. Fluids* **4**, 100 (1961).

Author Index

Numbers in italic indicate the page on which the complete reference is listed. Numbers in parentheses are reference numbers and show that an author's work is referred to although his name is not cited in the text.

Subject Index

A

Absorbed energy, 9ff.
Absorption, 95, 101, 114, 154, 164ff., 175, 179, 192, 194
Acceleration, charged-particle, 1
Anisotropy, 2, 52ff., 77ff., 84, 89, 96, 106, 133
Anti-Hermitian tensor, 22ff., 97
Averages
  quantum-mechanical, 6ff.
  statistical, 6, 8, 15, 37, 131, 170

C

Coefficients
  absorption, 114
  damping, 22, 46, 65ff., 102, 105
  differential scattering, 154, 156, 158, 171, 174
  diffusion, 3, 140ff., 148
  dynamic friction, 3, 140ff., 148
  integral scattering, 157, 174
  linear relating, 13, 21
  proportionality, 26, 28, 69, 72
  total scattering, 157, 159, 161ff.
  wave transformation, 165ff, 174ff.
Collision integral, 3, 140
Collisions, 1, 42, 79, 95, 115, 127ff., 134, 138ff., 150
Conductivity, 45, 204ff.
Correlation functions, 6, 10ff., 32, 34, 38ff., 47, 55, 59ff., 68, 79, 83ff., 89ff., 108, 132ff., 171, 175, 180, 183, 186, 192, 194ff., 200ff.
  space-time, 6ff., 13, 90ff., 132, 142, 179, 194
Correlators, 72, 78ff., 119, 172
Cyclotron resonance, 95, 114

D

Damping, 22, 27, 46ff., 49ff., 53, 65ff., 97, 102, 104ff., 108, 117ff., 121, 127, 165
Damping decrement, 51, 127
Debye Radius, 36, 42, 55, 125, 129, 140, 152
Density, 2, 20, 26ff., 29, 31, 37, 39ff., 46, 50ff., 54, 56ff., 61ff., 65ff., 73ff., 77ff., 82ff., 85ff., 89ff., 106ff., 113, 123ff., 128, 133ff., 138, 153ff., 156ff., 159, 163ff., 171ff., 179ff., 186ff., 189, 191ff., 194ff., 200, 203
  charge, 2, 27, 37, 39ff., 46, 54, 56ff., 62, 65ff., 70ff., 74ff., 77, 123,
  current 2, 20, 26ff., 29ff., 32, 34, 37, 39ff., 54, 58ff., 62, 67ff., 71, 78, 106, 123ff., 153, 179, 200
  electron 2, 31, 37, 50, 54, 61, 66, 68ff., 73, 78, 82ff., 90ff., 128, 134, 153ff., 156ff., 163ff., 168ff., 172ff., 187, 189
  energy flux, 156, 165ff., 171
  ion, 61, 66, 68ff., 71, 73, 78
Dielectric permittivity, 2ff., 18ff., 28ff., 31ff., 40ff., 47ff., 50ff., 54ff., 58ff., 61ff., 65, 67, 69, 77, 81ff., 85, 89, 93ff., 110ff., 119ff., 124ff., 130, 134, 136, 143, 145, 152ff., 164, 169, 172, 183ff., 186ff., 202ff.
Diffusion, 3, 140, 142ff., 148
Dispersion, 2, 12, 19ff., 25, 28, 31, 36, 44, 46ff., 51ff., 65, 79, 83ff., 87, 95ff., 98ff., 108ff., 111ff., 115ff., 120, 125, 128, 144, 151ff., 164ff., 169, 172, 174, 177, 180, 185ff., 196, 200ff.